Volume 67

Tissue-Specific Vascular Endothelial Signals and Vector Targeting, Part A

Advances in Genetics, Volume 67

Serial Editors

Theodore Friedmann
University of California at San Diego, School of Medicine, USA
Jay C. Dunlap
Dartmouth Medical School, Hanover, NH, USA
Stephen F. Goodwin
University of Oxford, Oxford, UK

Volume 67

Tissue-Specific Vascular Endothelial Signals and Vector Targeting, Part A

Edited by

Renata Pasqualini and Wadih Arap

David H. Koch Center
The University of Texas M. D. Anderson Cancer Center
Houston, TX, USA

AMSTERDAM • BOSTON • HEIDELBERG • LONDON
NEW YORK • OXFORD • PARIS • SAN DIEGO
SAN FRANCISCO • SINGAPORE • SYDNEY • TOKYO

ELSEVIER
Academic Press is an imprint of Elsevier

Academic Press is an imprint of Elsevier

525 B Street, Suite 1900, San Diego, CA 92101-4495, USA
30 Corporate Drive, Suite 400, Burlington, MA 01803, USA
32 Jamestown Road, London, NW1 7BY, UK
Radarweg 29, POBox 211, 1000 AE Amsterdam, The Netherlands

First edition 2009

Notice

ISBN: 978-0-12-375010-5
ISSN: 0065-2660

Printed and bound in USA

09 10 11 12   10 9 8 7 6 5 4 3 2 1

Working together to grow
libraries in developing countries

www.elsevier.com | www.bookaid.org | www.sabre.org

ELSEVIER   BOOK AID   Sabre Foundation
             International

Contents

Contributors

Numbers in parentheses indicate the pages on which the authors' contributions begin.

Wadih Arap (103) David H. Koch Center, The University of Texas M. D. Anderson Cancer Center, Houston, Texas 77030, USA

Joseph M. Backer (1) SibTech, Inc., Brookfield, Connecticut 06804, USA

Marina V. Backer (1) SibTech, Inc., Brookfield, Connecticut 06804, USA

Wouter H. P. Driessen (103) David H. Koch Center, The University of Texas M. D. Anderson Cancer Center, Houston, Texas 77030, USA

Carl V. Hamby (1) Department of Microbiology and Immunology, New York Medical College, Valhalla, New York 10595, USA

Mikhail G. Kolonin (61) The Brown Foundation Institute of Molecular Medicine for the Prevention of Human Disease, The University of Texas Health Science Center at Houston, Houston, Texas 77030, USA

Stefan Michelfelder (29) Department of Oncology and Hematology, Hubertus Wald Cancer Center, University Medical Center Hamburg-Eppendorf, Martinistrasse 52, D-20246 Hamburg, Germany

Michael G. Ozawa (103) David H. Koch Center, The University of Texas M. D. Anderson Cancer Center, Houston, Texas 77030, USA

Renata Pasqualini (103) David H. Koch Center, The University of Texas M. D. Anderson Cancer Center, Houston, Texas 77030, USA

Martin Trepel (29) Department of Oncology and Hematology, Hubertus Wald Cancer Center, University Medical Center Hamburg-Eppendorf, Martinistrasse 52, D-20246 Hamburg, Germany

1 Inhibition of Vascular Endothelial Growth Factor Receptor Signaling in Angiogenic Tumor Vasculature

Marina V. Backer,* Carl V. Hamby,[†] and Joseph M. Backer*

*SibTech, Inc., Brookfield, Connecticut 06804, USA
[†]Department of Microbiology and Immunology, New York Medical College, Valhalla, New York 10595, USA

ABSTRACT

Neovascularization takes place in a large number of pathologies, including cancer. Significant effort has been invested in the development of agents that can inhibit this process, and an increasing number of such agents, known as antiangiogenic drugs, are entering clinical trials or being approved for clinical use. The key players

Advances in Genetics, Vol. 67   0065-2660/09 $35.00
Copyright 2009, Elsevier Inc. All rights reserved.   DOI: 10.1016/S0065-2660(09)67001-2

involved in the development and maintenance of tumor neovasculature are vascular endothelial growth factor (VEGF) and its receptors (VEGFRs), and therefore VEGF/VEGFR signaling pathways have been a focus of anticancer therapies for several decades. This review focuses on two main approaches designed to selectively target VEGFRs, inhibiting VEGFR with small molecule inhibitors of receptor tyrosine kinase activity and inhibiting the binding of VEGF to VEGFRs with specific antibodies or soluble decoy VEGF receptors. The major problem with these strategies is that they appeared to be effective only in relatively small and unpredictable subsets of patients. An alternative approach would be to subvert VEGFR for intracellular delivery of cytotoxic molecules. We describe here one such molecule, SLT–VEGF, a fusion protein containing $VEGF_{121}$ and the highly cytotoxic catalytic subunit of Shiga-like toxin.   © 2009, Elsevier Inc.

I. INTRODUCTION

Growth of primary tumor and metastatic lesions beyond a few millimeters requires neovascularization that combines angiogenesis and vasculogenesis. In angiogenesis, endothelial cells of existing blood vessels undergo a complex process of reshaping, migration, growth, and organization into new vessels (Folkman, 1995). In vasculogenesis, endothelial progenitor cells migrate from the bone marrow to sites of angiogenesis and contribute significantly to the growth of new blood vessels (Rafii et al., 2002). Under normal circumstances, neovascularization, widely known under the name "angiogenesis," occurs during embryonic development, wound healing, and development of the corpus luteum. However, neovascularization takes place in a large number of pathologies, such as solid tumor growth, various eye diseases, chronic inflammatory states, and ischemic injuries. Therefore, significant research effort has been invested in the development of agents that can inhibit neovascularization, commonly known as antiangiogenic inhibitors (reviewed in Bergers and Hanahan, 2008; Gourley and Williamson, 2000; Jubb et al., 2006; Manley et al., 2004; Sledge and Miller, 2002; Thorpe et al., 2003). The first blockbuster drugs targeting VEGFR have already been approved by Food and Drug Administration (FDA) for treatment of several cancers with ~ 275,000 new US cases per year (Bergers and Hanahan, 2008; Chu, 2009; Izzedine et al., 2009; Jubb et al., 2006; Ruan et al., 2009). The potential of these drugs is enormous, as judged by over 230 US-registered Phase III clinical trials for all major cancers with an estimated 12 million new cases annually, worldwide (Hayden, 2009). This review will focus on strategies designed to selectively target the key players involved in the development and maintenance of tumor neovasculature: vascular endothelial growth factor (VEGF) and its receptors (VEGFRs).

II. VEGF/VEGFR SIGNALING PATHWAY AS A TARGET OF ANTIANGIOGENIC THERAPY

A. VEGF is a critical positive regulator of angiogenesis

Several positive and negative regulators control the process of angiogenesis. It is hypothesized that the shift in equilibrium between these regulators, known as the "angiogenic switch," is responsible for angiogenesis in pathological situations (Hanahan and Folkman, 1996). The crucial positive regulator of angiogenesis is VEGF-A, also known as vascular permeability factor (Ferrara, 2009). VEGF-A is a secreted dimeric glycoprotein produced by many cells. VEGF is a potent angiogenic factor *in vivo* and induces numerous responses in endothelial cells in tissue culture. There are at least three more members of VEGF family: VEGF-B, -C, and -D (Ferrara, 2009; Lohela *et al.*, 2009). VEGF-B has a very limited angiogenic potential, and is involved in regulating lipid metabolism in the heart. VEGF-C and VEGF-D induce lymphangiogenesis and have been implicated in stimulating metastasis.

Four different forms of human VEGF (VEGF-A), containing 121, 165, 189, and 206 amino acid residues, arise from alternative splicing of mRNA. The first three forms are common in adult organisms while $VEGF_{206}$ is expressed during embryonic development. $VEGF_{121}$ is a circulating form of the growth factor. $VEGF_{165}$, $VEGF_{189}$, and $VEGF_{206}$ contain heparin-binding domain(s) in the C-terminal portion. Interactions with heparin-containing extracellular proteoglycans lead to the deposition of $VEGF_{165}$ and particularly $VEGF_{189}$ and $VEGF_{206}$ in the extracellular matrix. VEGF is expressed by normal and tumor cells and the control of VEGF expression appears to be regulated on several levels (Claffey and Robinson, 1996). Expression of VEGF is upregulated in response to hypoxia and nutritional deprivation, suggesting a feedback loop between tumor growth and the ability of tumor cells to induce host angiogenic responses (Veikkola and Alitalo, 1999).

B. VEGF receptors

VEGFs and their endothelial tyrosine kinase receptors are central regulators of vasculogenesis, angiogenesis, and lymphangiogenesis (Lohela *et al.*, 2009). VEGF signaling through VEGFR-2 is the key process in angiogenesis, and inhibition of VEGF/VEGFR-2 signaling is the core of antiangiogenic strategy for cancer therapy. VEGFR-1 acts mostly as a negative regulator of VEGF-mediated angiogenesis during development, and as a stimulator of pathological angiogenesis when activated by its specific ligands PlGF (placenta-derived growth factor) and VEGF-B. VEGFR-3 is a key player in lymphangiogenesis, and contributes to control of angiogenic sprouting angiogenesis, acting together with VEGF/VEGFR-2.

VEGFR-2 receptor is a single-span transmembrane protein tyrosine kinase expressed predominantly in endothelial cells. VEGFR-2 belongs to the immunoglobulin superfamily. It contains seven Ig-like loops in the extracellular domain and shares homology with the receptor for platelet-derived growth factor (Terman and Dougher-Vermazen, 1996; Veikkola et al., 2000). VEGF binding to VEGFR-2 induces receptor dimerization followed by tyrosine phosphorylation of the SH2 and SH3 domains in the dimer. Tyrosine phosphorylation activates signal transduction pathways, which leads to calcium mobilization, activation of phospholipases C and D, polymerization of actin, changes in cell shape and chemotactic and mitogenic responses. VEGFR-2/VEGF complex is internalized via receptor-mediated endocytosis (Bikfalvi et al., 1991).

Immunohistochemical analysis indicated that endothelial cells at the sites of angiogenesis express significantly higher numbers of VEGFR-2 than quiescent endothelial cells (Brown et al., 1995; Couffinhal et al., 1997; Koukourakis et al., 2000). Recent data on molecular imaging of VEGFR in tumors support these observations (Backer et al., 2005, 2007; Blankenberg et al., 2004; Cai et al., 2006, Hsu et al., 2007; Levashova et al., 2008; Wang et al., 2007, 2009). This difference in VEGFR prevalence presents new opportunities for selective targeting of endothelial cells at the sites of angiogenesis.

The fundamental problem in development of antiangiogenic therapeutics is finding targets that can differentiate between the relatively small number of tumor endothelial cells and the very large ($\sim 10^{12}$) number of normal endothelial cells in the body; and the VEGF/VEGFR signaling pathway appears to be such a target. A significant number of experimental therapeutics targeting VEGF/VEGFR signaling in tumor vasculature have been tested for all major cancers, and multiple late stage clinical trials for some of these drugs are in progress (Bergers and Hanahan, 2008; Chu, 2009; Izzedine et al., 2009; Jubb et al., 2006; Ruan et al., 2009).

III. STRATEGIES TO INHIBIT THE VEGF/VEGFR SIGNALING

A. Blocking VEGF/VEGFR binding

VEGF- and VEGFR-specific neutralizing antibodies or soluble VEGFR-based traps that prevent binding of VEGF to its receptors present the first antiangiogenic strategy. This strategy is based on the assumption that continuous depravation of VEGF/VEGFR signaling is more detrimental for tumor endothelial cells than it is for normal endothelial cells. The most effective VEGF-specific neutralizing antibody developed so far, bevacizumab (Roche/Genentech, trade name Avastin), is the first FDA-approved drug targeting VEGF/VEGFR signaling pathway. Bevacizumab is a humanized murine monoclonal antibody that binds human VEGF and, therefore, diminishes VEGFR signaling, which is presumably more important for growth and maintenance of tumor endothelium than for normal endothelium. It is approved for treatment of metastatic colorectal cancer in combination with chemotherapy, either with oxaliplatin/5-FU/leucovorin (FOL-FOX4) therapy as first-line treatment, or with 5-fluorouracil-based therapy as second-line treatment. It is also approved for treatment of nonsquamous nonsmall cell lung cancer in combination with carboplatin and paclitaxel, for metastatic breast cancer in combination with paclitaxel, and as a single agent for treatment of progressing glioblastoma following prior therapy. A recent review by Chu (2009) summarizes the advantages of VEGF neutralization strategy that involves either VEGF- and VEGFR-specific antibodies or decoy receptors for VEGF (VEGF-traps).

B. Inhibition of VEGF-induced signaling

This strategy is based on two assumptions: (1) efficacious concentrations of inhibitors can be created in tumor endothelial cells overexpressing VEGFR-2 and (2) continuous inhibition of VEGFR-2-mediated signal transduction is more detrimental for tumor endothelial cells than it is for normal endothelial cells. Over the last decade, there has been a dramatic increase in the number of VEGFR-2 inhibitors that demonstrated successful inhibition of tumor growth in preclinical testing. Two inhibitors, sunitinib and sorafenib, are already approved for clinical use and the number of VEGFR tyrosine kinase inhibitors (TKIs) entering clinical trial as anti-cancer agents is rapidly increasing (Johannsen et al., 2009; van Cruijsen et al., 2009).

The majority of known kinase inhibitors bind to the ATP-binding site and therefore most of them act as multikinase inhibitors. Additional targets may result both in an increased efficacy and an increased toxicity of these drugs (van Cruijsen et al., 2009). So far, the most specific TKI act at least on these three major types of VEGFRs: VEGFR-1 (Flt-1), VEGFR-2 (Flk-1/KDR), and VEGFR-3 (Flt-4). Currently, significant effort is being invested in the development of more specific TKIs (Peifer et al., 2009; Schmidt et al., 2008; Srivastava et al., 2009) as well as exploring the potential of dual- and multitargeted TKIs (Pennell and Lynch, 2009).

1. VEGFR-specific and broad-spectrum TKIs induce rapid regression of tumor vasculature

Despite the fact that increasing numbers of antiangiogenic drugs are entering clinical trials or being approved for clinical use, molecular mechanism(s) of action of VEGF/VEGFR-targeting TKIs are not fully characterized and understood (Bergers and Hanahan, 2008; Ellis and Hicklin, 2008). In general, VEGFR-specific as well as broad-spectrum TKIs rapidly and efficiently induce regression of tumor vasculature (Chang *et al.*, 2007; Mancuso *et al.*, 2006; Mendel *et al.*, 2003; Palmowski *et al.*, 2008; Smith *et al.*, 2007). Experiments on spontaneous RIP-Tag2 tumors and implanted Lewis lung carcinomas in mice demonstrated that inhibition of VEGFR signaling by two TKIs, AG-013736 or AG-028262, led to a significant regression of tumor vasculature after 7 days of treatment, resulting in empty sleeves of basement membrane partially covered with pericytes and smooth muscle cells (Mancuso *et al.*, 2006). Using high-frequency power contrast-enhanced Doppler ultrasound imaging, Palmowski and colleagues were able to register the functional effects of sunitinib on tumor vascularization in human epidermoid carcinoma A431 xenografts as soon as few hours after the start of the treatment (Palmowski *et al.*, 2008), with a significant decrease in VEGFR-2-specific staining of endothelial (CD31-positive) cells after 2–3 days of the treatment. A similar rapid regression of tumor vasculature in different murine tumor xenograft models has also been reported for other VEGFR-2 TKIs, such as sorafenib (Chang *et al.*, 2007) and AZD2171 (Bozec *et al.*, 2008).

To monitor changes in VEGFR prevalence in the course of treatment with antiangiogenic drugs, we developed a family of molecular tracers for multimodality imaging that are based on an engineered single-chain (sc) VEGF (Backer *et al.*, 2007). scVEGF-based imaging tracers for positron emission tomography (PET), single photon emission computed tomography (SPECT), and near infrared fluorescent (NIRF) imaging selectively accumulate in tumor endothelial cells expressing high levels of VEGFR-2 via VEGFR-2-mediated tracer uptake. This accumulation is particularly prominent in the angiogenic rim area at the edges of the growing tumors where active angiogenesis takes place (Backer *et al.*, 2007; Levashova *et al.*, 2008). We reasoned that imaging changes in VEGFR prevalence in response to TKI can provide rapid and noninvasive assessment of drug effects, and tested this hypothesis with FDA-approved sunitinib and experimental therapeutic pazopanib, a TKI-targeting VEGFR, PDGFR, and c-Kit, currently under clinical development (Kumar *et al.*, 2007; Sloan and Scheinfeld, 2008). Our SPECT imaging with scVEGF/^{99m}Tc of human colon cancer HT29 tumors grown in Swiss nude mice indicated a significant decrease in tracer uptake after 5 days of pazopanib treatment, with a remarkable pazopanib-induced depletion of CD31$^+$/VEGFR-2$^+$ endothelial cells in tumor vasculature that occur at the early stages of pazopanib treatment (Blankenberg *et al.*, 2010).

2. Posttreatment vascular resurgence and associated adverse effects of TKI treatment

Aggressive regrowth of tumor blood vessels shortly after the termination of TKI treatment, or after prolonged treatment with antiangiogenic drugs, is also well documented. Mancuso *et al.* (2006) reported that 1 day after drug withdrawal, endothelial sprouts started to grow into the scaffold provided by empty sleeves of basement membrane, resulting in fully revascularized tumors with completely restored vessel patency by 7 days. In our work, robust revascularization at tumor edges 2–3 days after withdrawal of sunitinib or after 15-day treatment with pazopanib was detected by SPECT imaging with scVEGF/^{99m}Tc tracer and confirmed by autoradiography and immunohistochemistry (Blankenberg et al., 2010). Rapid vascular resurgence in the course of prolonged treatment with VEGFR-specific TKIs has been implicated in the resistance to these drugs developed by patients (Dempke and Heinemann, 2009; Johannsen et al., 2009). Molecular mechanisms of resistance identified to date include (1) upregulation of bFGF signaling, (2) overexpression of MMP-9, (3) increased levels of SDF-1alpha, and (4) HIF-1alpha-induced recruitment of bone marrow-derived CD45$^+$ myeloid cells (Dempke and Heinemann, 2009).

The mechanisms of resistance and, in general, the mechanisms that determine the response of individual patients to drugs targeting VEGF/VEGFR signaling are complex and the outcome of the treatment is unpredictable (Bergers and Hanahan, 2008; Ellis and Hicklin, 2008; Jain, 2005; Jain et al., 2006; Lin and Sessa, 2004; Siemann et al., 2005). Indeed, the first TKIs approved by FDA for clinical use, sunitinib and sorafenib, appeared to be effective only in relatively small and unpredictable subsets of patients, while the treatment could result in serious side effects (Bergers and Hanahan, 2008; Jubb et al., 2006). In part, it most likely reflects the redundancy of angiogenic pathways that allows continuous growth of tumor vasculature even when signaling by one of the positive regulators is gradually inhibited, as well as difficulties in reaching efficacious concentrations without significant side effects to normal endothelium (Saaristo et al., 2000). The major reason of the observed adverse effects of antiangiogenic therapy is thought to be the inhibition of the biological function of endothelial cells in healthy tissue (Manley et al., 2004; van Cruijsen et al., 2009). Proteinuria and hypertension have been reported as the most frequent side effects of antiangiogenic therapies (Grunwald et al., 2009; Izzedine et al., 2009); acute kidney injury has also been reported (Gurevich and Perazella, 2009).

There is also a controversy regarding a posttreatment recovery and regrowth of endothelial cells. On one hand, it might provide for better access of chemotherapeutic agents to tumor cells, supporting combination therapy and, particularly, metronomic combinations (Jain, 2005; Kerbel and Kamen, 2004). On the other hand, posttreatment vascular resurgence can stimulate tumor invasiveness and metastatic dissemination (Ebos et al., 2009; Loges et al., 2009; Paez-Ribes et al., 2009). Acceleration of metastasis was observed in mouse tumor models after short-term therapy with sunitinib, suggesting possible "metastatic conditioning" in multiple organs. Also, the inhibitors targeting the VEGF/VEGFR pathway that demonstrated antitumor effects in mouse models of pancreatic neuroendocrine carcinoma and glioblastoma concomitantly elicit tumor progression to greater malignancy and invasiveness and increased lymphatic and distant metastases (Paez-Ribes *et al.*, 2009). Importantly, these observations of metastatic acceleration were in contrast to well-documented antitumor benefits obtained with the same type cancer cells in animal models.

In attempt to reduce lymphatic metastasis, Padera *et al.* (2008) used a combination of two different TKIs, cediranib and vandetanib. The inhibitory effects of cediranib on VEGFR-3-mediated endothelial cell function and lymphangiogenesis were demonstrated independently (Heckman *et al.*, 2008). Cediranib blocked VEGF-induced proliferation, survival, and migration of lymphatic and blood vascular endothelial cells *in vitro. In vivo*, cediranib prevented angiogenesis and lymphangiogenesis induced by VEGF-expressing adenoviruses and compromised the blood and lymphatic vasculatures of VEGF-C-expressing tumors (Heckman *et al.*, 2008). When given in combination with vandetanib during tumor growth, cediranib reduced the diameters of the draining lymphatic vessels, the number of tumor cells arriving in the draining lymph node, and the incidence of lymphatic metastasis. However, neither agent was able to prevent lymphatic metastasis when given after tumor cells had seeded the lymph node (Padera *et al.*, 2008).

To summarize, the overall outcome of VEGFR-2 inhibition with TKIs appears to be transient; the regression of tumor vasculature is usually followed by revascularization, particularly when the treatment is interrupted. It is not clear why initially tumor endothelial cells are sensitive to inhibition of VEGF/VEGFR signaling, while normal endothelial cells are not, and why, eventually, this sensitivity disappears. Furthermore, new data suggest that inhibition of VEGF/VEGFR signaling might result in more aggressive tumors.

IV. USING VEGFR-2 FOR TARGETED DELIVERY OF POWERFUL TOXIC AGENTS

An alternative approach to destruction of tumor endothelium would be targeted delivery of highly toxic agents, such as potent bacterial or plant toxins, to tumor endothelial cells. The idea of using bacterial or plant toxins for selected destruction of unwanted cells is not new. Development of engineered chimeric toxins, so-called immunotoxins, for targeted delivery of toxic moieties to cancer cells has been explored for over 30 years (Holzman, 2009; Kreitman, 2006, 2009; Pastan *et al.*, 2006, 2008). In this approach, DNA encoding a targeting moiety (a tissue-specific antibody or a growth factor) is cloned in-frame with a potent toxin and expressed in bacteria, so that the resulting immunotoxin contains the targeting and toxin moieties in a single polypeptide. Alternatively, the toxin is chemically conjugated to a targeting protein in a way that preserves activity of both toxin and targeting moiety.

The accumulated experience with immunotoxins suggests that VEGF–toxin fusion proteins might be useful for inhibition of angiogenesis. The advantages of targeting toxins to VEGFR-2 in tumor angiogenesis are as follows: (1) VEGFR-2 is available directly from the bloodstream; and therefore a VEGF–toxin fusion protein does not need to extravasate from blood vessels and penetrate into tumor tissue; (2) VEGFR-2 is overexpressed on tumor endothelial cells relative to normal endothelium, providing a measure of protection against systemic toxicity; (3) VEGF is rapidly internalized via VEGFR-2-mediated endocytosis, providing a mechanism for efficient accumulation of VEGF–toxin in endothelial cells; (4) depletion of VEGFR-2 overexpressing cells in tumor vasculature might be achieved before the onset of a host immune response and systemic side effects.

A. Plant and bacterial toxins for targeting tumor endothelial cells overexpressing VEGFR-2

1. Diphteria toxin

Diphteria toxin (DT) is the prototype for the family of ADP ribosylating toxins (Deng and Barbieri, 2008). The N-terminal domain represents the enzymatically active A domain, and the C-terminal domain comprises a translocation (T) domain that facilitates translocation of the A domain into the cytoplasm of the host cells and a receptor-binding (R) domain that binds the toxin to host cell surface receptors. Activities of all three domains, A–T–R, are required for the observed toxicity of DT. Receptor binding triggers the entry of DT into the lumen of a developing endosome by receptor-mediated endocytosis. Upon acidification, T domain facilitates the translocation of the A domain across the endosomal membrane and into the host cell cytoplasm, where the A domain catalyzes the ADP-ribosylation of eukaryotic elongation factor-2 (eEF-2) to inhibit protein synthesis, thus leading to cell death (Deng and Barbieri, 2008).

   Two groups constructed targeted toxins containing VEGF and either chemically conjugated, or genetically fused catalytically active A–T domains of DT. These proteins displayed selective cytotoxicity against human endothelial cells (HUVEC) and Kaposi's sarcoma cells expressing VEGFR-2, and also suppressed angiogenesis in the chick chorioallantoic membrane model and growth of tumors in immunodeficient mice (Arora et al., 1999; Olson et al., 1997; Ramakrishnan et al., 1996). Although these experiments provided a "proof-of-principle" that VEGF–toxin conjugates or fusion proteins may work in vivo, further development of DT–VEGF constructs is doubtful for several reasons. First, cytotoxicity to HUVEC indicates that DT–VEGF would be cytotoxic to normal endothelial cells expressing a similar to HUVEC level of VEGFR-2 ($\sim 25,000$/cell). Second, well-documented renal and liver toxicity of various

DT-containing fusion proteins (Frankel *et al.*, 2003; Vallera *et al.*, 1997) would present serious obstacles for clinical development of VEGF–DT fusion protein. Finally, widespread preexisting immunity against DT is expected to decrease activity of any DT-based immunotoxin. In fact, these problems were encountered in the development of DT fused to another growth factor, epidermal growth factor (EGF). An EGF–DT fusion protein significantly inhibited tumor growth in animal models, up to complete tumor regression (Liu *et al.*, 2003) and reached testing in Phase I/II clinical trials. However, the preexisting DT-specific immunity and high liver toxicity (Cohen *et al.*, 2003; Foss *et al.*, 1998; Liu *et al.*, 2003) prevented further clinical development of EGF–DT. The preexisting immunity to DT-based immunotoxins is not surprising, given that DT toxoid is a part of DTaP (diphtheria–tetanus–pertussis) vaccine that is currently approved and, in a slightly modified composition, has been widely used for massive vaccination in the past. The DT component of the vaccine provides for reportedly high level of immunization and a long-memory immune response. Indeed, one-fourth of the breast cancer patients involved in EGF–DT Phase I/II clinical trials had high pretreatment DT antibody titers, and all patients had high anti-DT antibody titers 1 month posttreatment (Theodoulou *et al.*, 1995). Formation of antibody–DT complexes may be responsible, in part, for reduced tumor accumulation, rapid clearance of EGF–DT cytotoxins from the circulation, and hepatotoxicity, which is usually attributed to the binding of basic residues of the Fv antibody region to negatively charged hepatic cells (Kreitman, 2006).

2. Gelonin

Gelonin is a member of ribosome-inactivating protein (RIP) family (Stirpe *et al.*, 1980). RIPs are enzymes which depurinate rRNA and other polynucleotide substrates, thus inhibiting protein synthesis. However, the biological activity of RIPs is not fully understood, as it is sometimes independent from the inhibition of protein synthesis (Stirpe and Battelli, 2006).

A VEGF–gelonin fusion protein, VEGF(121)/rGel, inhibits growth of endothelial cells in tissue culture and growth of subcutaneous tumors in several mouse models, including glioma, melanoma, prostate carcinoma, and pulmonary metastases of MDA-MB-231 breast tumors (Hsu *et al.*, 2007; Ran *et al.*, 2005; Veenendaal *et al.*, 2002). Interestingly, VEGF(121)/rGel works through the VEGFR-2 receptor, while endothelial cells overexpressing VEGFR-1 were not sensitive to this fusion protein (Ran *et al.*, 2005; Veenendaal *et al.*, 2002). In a somewhat contradictory finding, VEGF(121)/rGel inhibits growth of VEGFR-1 positive osteoclast precursor cells (Mohamedali *et al.*, 2006), which may play an important role in suppression of skeletal osteolytic lesions. Indeed, systemic treatment of nude mice bearing intrafemoral prostate PC3 tumors with VEGF

(121)/rGel resulted in complete regression of bone tumors in 50% of treated animals, with no development of lytic bone lesions (Mohamedali et al., 2006). Immunohistochemical analysis confirmed that VEGF(121)/rGel suppressed tumor-mediated osteoclastogenesis in vivo.

Notably, gelonin does not appear to generate vascular leak syndrome that limits the use of other toxins, including DT (Talpaz et al., 2003), and, from this perspective, VEGF(121)/rGel may be a promising agent for targeting VEGF receptors in tumor vasculature. However, the reported low yield of VEGF(121)/rGel produced in Escherichia coli (0.23 mg/l) is problematic for its preclinical development (Veenendaal et al., 2002).

3. Shiga-like toxin

Shiga-like toxin (SLT) holotoxin combining cell-receptor-binding B subunits and catalytic A subunits is considered a "natural killer" for endothelial cells. This is due to binding of SLT B subunit to the cell surface receptor, globotrioaosylceramide $Gb_3/CD77$, expressed on endothelial cells and enhanced sensitivity of endothelial cells to enzymatic activity of A subunit. In fact, damage to endothelial cells caused by SLT plays a causative role in the pathogenesis of hemorrhagic colitis and hemolytic uremic syndrome induced by toxin-producing E. coli O157: H7 (Kaplan et al., 1990; Obrig et al., 1987, 1993; Richardson et al., 1988).

SLT is composed of a single 32 kDa A subunit associated with a ring-shaped pentamer of 7 kDa receptor-binding B subunits. After binding to the cell surface receptor $Gb_3/CD77$, SLT is endocytosed via clathrin-mediated pathway (Sandvig et al., 1989) and transported to the trans-Golgi network. In the trans-Golgi the endoprotease furin cleaves the A subunit into A_1 (27.5 kDa) and A_2 (4.5 kDa) fragments (Brigotti et al., 1997; Olsnes et al., 1981). The released A_1 fragment is a specific N-glycosidase that cleaves off a single adenine residue in position 4324 in $5'$ terminus of 28 S rRNA of the 60 S ribosome subunit (Saxena et al., 1989).

The cleavage of A_{4324} from 28 S rRNA inhibits binding of the elongation factor (eEF-1)/aminocyl-tRNA complex to ribosomes, resulting in inhibition of the protein synthesis and cell death. There is evidence that bacterial toxins affecting activity of elongation factors may induce apoptosis not only through a shortage of the de novo synthesized proteins but also through other stress-response mechanisms (Duttaroy et al., 1998; Iordanov and Magun, 1999; Iordanov et al., 1997). This high efficiency of SLT-1 makes it an attractive candidate for a therapeutic application when coupled to a highly selective vehicle (Williams et al., 1997). Indeed, Al-Jaufy et al. successfully targeted human HIV-1 infected cells by fusing truncated A subunit of SLT-1 to the N-terminus of CD4. These fusion proteins entered infected cells via the interaction between the CD4 moiety and the HIV-1 gp120–gp41 complex (Al-Jaufy et al., 1994, 1995).

B. Development of SLT–VEGF chimeric toxin for targeting VEGFR-2 overexpressing cells in tumor neovasculature

Since SLT is a "natural killer" of endothelial cells, it seems to be well suited for inhibition of angiogenesis. In order to target SLT to endothelial cells, we constructed a fusion protein that contains the catalytic A subunit of SLT and VEGF moieties. A 121 aa form of VEGF was chosen for this construct since it lacks a heparin binding domain found in the longer forms of VEGF, and therefore displays lower nonspecific binding to the cell-surface and extracellular matrix heparan sulfate proteoglycans.

Although SLT is almost three times larger than $VEGF_{121}$, we found earlier that the presence of large N-terminal extensions is well tolerated by VEGF (Backer and Backer, 2001a). We expressed SLT–VEGF in *E. coli* and found that both SLT and VEGF moieties retain their functional activities (Backer and Backer, 2001b; Backer *et al.*, 2001). SLT–VEGF fusion protein inhibited growth of porcine aortic endothelial (PAE) cells engineered to express $\sim 2.5 \times 10^5$ VEGFR-2/cell with $IC_{50} \sim 0.1$ nM. We also engineered a panel of cells expressing VEGFR-2 at the levels from 25,000 to 2,500,000/cell and found that the IC_{50} values are inversely proportional to the levels of VEGFR-2 expression. The selective toxicity of SLT–VEGF to growing endothelial cells was also demonstrated by the fact that confluent PAE/KDR cells were not sensitive to this protein. In contrast, growing PAE/KDR cells treated with SLT–VEGF underwent apoptosis and cytolysis (Backer and Backer, 2001b). These results indicated that SLT–VEGF might display a desired selective cytotoxicity for growing endothelial cells overexpressing VEGFR-2 *in vivo*.

1. Optimization of SLT–VEGF fusion protein for preclinical testing

Initially we have constructed, expressed, and purified an SLT–VEGF fusion protein that was cloned into a plasmid with an ampicillin resistance marker, contained His- and S-affinity tags, and was purified with a yield of ~ 0.1 mg/l (Backer and Backer, 2001b). Since these features are not compatible with clinical development of biopharmaceuticals, we have redesigned the SLT–VEGF fusion construct to an "FDA-friendly" format. The current version of SLT–VEGF is cloned into the pET29 bacterial expression vector carrying a kanamycin resistance marker and does not contain affinity tags. Like the previous His- and S-tagged version, new SLT–VEGF protein was found in inclusion bodies and required refolding to maximize formation of the functionally active dimeric form. Interestingly, the refolding procedure previously developed for tagged SLT–VEGF yielded mostly functionally inactive monomers (>90%) for the tag-less SLT–VEGF (not shown). Therefore, we introduced additional steps in the refolding procedure aimed specifically at disrupting scrambled disulfide bonds in solubilized SLT–VEGF. The additional steps included an 18-h treatment with 10 mM DTT at 4 °C in the presence of the sulfonating agents, Na_2SO_3 and $Na_2S_4O_6$, followed by an 18-h treatment with another reducing agent, TCEP, at a final concentration of 5 mM. After DTT/TCEP treatment, the protein was refolded via slow dialysis under Red-Ox conditions in the presence of reduced and oxidized glutathione and purified by anion exchange chromatography, yielding 3–5 mg/l of purified protein. Although still moderate, this yield is 10–20 times higher than that reported for two other VEGF–toxin fusion proteins, VEGF–DT (Arora *et al.*, 1999) and VEGF–gelonin (Veenendaal *et al.*, 2002).

2. Dose/route selection for SLT–VEGF administration

Drug dosage and administration route were optimized using nontumor-bearing Balb/c mice ($n = 5$) that received either intravenous (i.v.) or subcutaneous (s.c.) injections of varying amounts of SLT–VEGF (Table 1.1). A single i.v. injection of SLT–VEGF at 2.25 mg/kg (45 μg/mouse) was extremely toxic, killing 80% of mice, while a single s.c. injection of the same dose resulted in a transient morbidity followed by full recovery in 3–5 days. Five i.v. injections of SLT–VEGF at the level of 0.25 mg/kg (5 μg/mouse) killed 40% of mice, while all mice given five s.c. injections on the same schedule at the same dose survived. Based on these experiments, a regimen of multiple s.c. SLT–VEGF injections at 0.05 mg/kg/ injection (1 μg/mouse/injection) was selected for treating tumor-bearing mice. It should be noted that after two to three s.c. injections at 1 μg/mouse/injection, mice appeared to experience transient discomfort. However, all mice recovered within a few days and additional injections were not associated with any noticeable adverse effects. Furthermore, seven s.c. injections (0.05 mg/kg/injection) over 20 days did not affect animal weight (21.2 ± 1.1 g for treated and 21.3 ± 1.0 g for control mice, $n = 15$) and did not result in detectable histopathology of major organs (data not shown).

Table 1.1. Route and Dose Optimization for SLT–VEGF Injections

SLT–VEGF per injection (μg)	Number of injections	Type of injection	
		Intravenous	Subcutaneous
45	1	80% dead	Sick
5	5	40% dead	Normal

3. Immunogenicity in mice

Since the SLT moiety is of bacterial origin, SLT–VEGF may be strongly immunogenic and it might hamper its development as a therapeutic. The immunogenicity of SLT–VEGF in Balb/c mice-bearing 4T1 tumors was tested in groups ($n = 5$) receiving a single ($1\times$), or either three ($3\times$) or five ($5\times$) biweekly s.c. SLT–VEGF injections of 0.05 mg/kg/injection. Control mice received s.c. injections of equivalent volumes of PBS. Sera were collected on day 5 after the last injection. Anticipating that human VEGF would induce an immune response in mice, we first removed anti-VEGF antibodies from the sera. Saturating amounts of human recombinant Hu-tagged VEGF were added to each serum sample, and VEGF–antibody complexes were allowed to form for 16 h at 4 °C. Both VEGF–antibody complexes and unbound VEGF were then removed from solution by capture with S-protein agarose that binds Hu-tag with high affinity (Backer *et al.*, 2003). VEGF-depleted sera were tested for binding to SLT–VEGF immobilized on the surface of 96-well plates. We found that a single injection of SLT–VEGF was not sufficient to elicit an anti-SLT response, while three and five injection schedules resulted in a relatively low-titered (1:3000–1:12,000) anti-SLT response (Fig. 1.1A). Interestingly, antibodies developed in four out of five mice receiving three injections were capable of protecting 293/KDR cells from cytotoxic activity of SLT–VEGF (Fig. 1.1B). Since the sera were depleted of anti-VEGF-specific antibodies, the data demonstrated that these mice developed neutralizing anti-SLT antibodies.

4. SLT–VEGF-induced tumor growth inhibition in mouse tumor models

a. Prostate cancer model

PC3 tumors were established by inoculating 5×10^6 cells/animal s.c. in the backs of Ncr nu/nu mice. Starting on day 7, mice ($n = 15$) received seven biweekly s.c. injections of SLT–VEGF at a dose of 0.05 mg/kg/injection (1 μg/mouse/injection). This SLT–VEGF treatment significantly inhibited PC3 tumor growth (Fig. 1.2A) by 30 days of treatment. Tumor growth inhibition was also observed with five s.c. injections at the same dose (Fig. 1.2B) by 25 days of treatment. However, the same cumulative dose of 5 μg/mouse, given as a single s.c. injection, did not affect overall tumor growth, although a lag in early tumor growth was observed (Fig. 1.2C). Taken together, these data suggested that effects of SLT–VEGF can be "saturated" by the finite number of injections, but not by a single large dose.

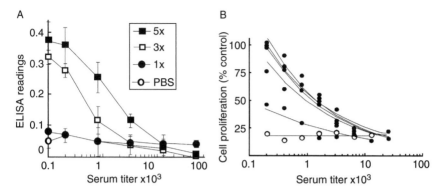

Figure 1.1. Multiple injections of SLT–VEGF induce a weak immune response in mice. (A) Sera were harvested from mice that received s.c. injections of SLT–VEGF on three different dosing schedules: one injection of 1 μg SLT–VEGF (1×), three injections of 1 μg SLT–VEGF (3×), and three injections of 1 μg plus two injections of 0.33 μg of SLT–VEGF (5×). Control mice received five PBS injections. The anti-VEGF-depleted sera were titrated by serial fourfold dilutions from an initial 1:200 dilution and reacted in 96-well Maxisorb plates coated with 1 μM SLT–VEGF. Bound mouse immunoglobulins were detected with HRP-conjugated antimouse IgG diluted 1:2000, developed by OPD staining and read at 405 nm. (B) VEGF-depleted sera of 3× injected mice neutralize the cytotoxic effect of SLT–VEGF on 293/KDR cells. Cells were plated in 96-well plates, 500–1000 cells/well 20 h before the experiment. The anti-VEGF-depleted sera of mice receiving three injections of 1 μg SLT–VEGF were titrated by serial twofold dilutions from an initial 1:100 dilution in complete culture medium containing 1 nM SLT–VEGF, and then added to cells in triplicate wells. Numbers of viable cells were determined after 96 h of incubation by an MTT-based assay. Black circles, sera of SLT–VEGF-treated mice (n = 5); open circles, serum of a PBS-injected control mouse.

b. Pancreatic cancer model

In experiments performed at Dr. Hotz's laboratory (Charité School of Medicine Campus Benjamin Franklin, Berlin, Germany), human pancreatic cancer cells AsPC-1 and HPAF-2 were orthotopically implanted in nude mice. Animals were randomized into control and treatment groups and administration of SLT–VEGF started either 3 days (prophylaxis) or 6 weeks (therapy) after tumor induction. Volume of the primary tumor, local infiltration, metastatic spread, and microvessel density were determined at autopsy after 14 weeks. Treatment with the SLT–VEGF fusion protein reduced tumor growth and metastatic spread in both prophylaxis and therapy groups, resulting in a significantly increased 14-week survival of the treated animals. Reduced microvessel density indicated that this effect was mainly due to the toxic effect of SLT–VEGF on endothelial cells of the tumor vasculature. Importantly, SLT–VEGF therapy was not associated with systemic side effects such as weight loss (Hotz et al., 2006).

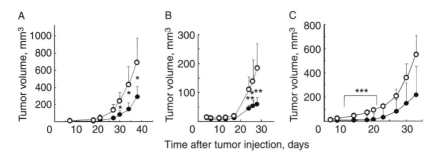

Figure 1.2. SLT–VEGF inhibits growth of PC3 tumors. PC3 human prostate cancer cells (ATCC # CRL-1435) were implanted s.c. in the middle/lower backs of 5–6-week-old NCr (nu/nu) mice, 5×10^6 cells/injection. SLT–VEGF s.c. injections were given between the ears at 3.5–4.5 cm distances from growing tumors. (A) Seven biweekly s.c. injections at 0.05 mg/kg/injection (1 μg/mouse/injection, $n = 15$). (B) Five biweekly s.c. injections at 0.05 mg/kg/injection (1 μg/mouse/injection, $n = 5$). (C) Single s.c. injection at 0.25 mg/kg/injection (5 μg/mouse/injection, $n = 5$). *$p \leq 0.001$, **$p \leq 0.01$, ***$p \leq 0.05$.

c. Mouse mammary carcinoma

There is always a concern that human tumor xenografts grown in immuno-deficient mice are quite artificial tumor models that fail to provide adequate information regarding tumor behavior in syngeneic hosts. Therefore, we tested 4T1 mouse mammary carcinoma, an aggressive rapidly growing tumor in syngeneic Balb/c mice. Tumors were established by s.c. injections of as low as 2000 cells/animal. Mice-bearing 4T1 tumors ($n = 8$–15) were treated with three different SLT–VEGF regimens starting on day 4 after tumor cell injection. Tumor growth was analyzed for every individual mouse. By plotting tumor growth rates for individual mice as linear regression curves, it was possible to clearly distinguish responder and nonresponder mice in each group (Fig. 1.3). Fisher's exact test was applied to compare the differences in the proportion of responder and nonresponder mice in each treatment group. We found that the proportion of mice whose tumors respond to SLT–VEGF treatment increased with an increasing cumulative dose of SLT–VEGF. However, the rate of tumor growth in responders did not depend on the cumulative dose, suggesting that effects of SLT–VEGF in responsive mice are readily saturated. In other words, tumor growth was inhibited to about the same extent in all responder mice regardless of the treatment regimen. This surprising finding suggests that pharmacodynamic endpoints of SLT–VEGF treatment are achieved by turning off rather than by gradual inhibition of targeted processes.

In a separate set of experiments we also tested whether SLT–VEGFci, a fusion protein containing a catalytically inactive mutant of SLT (described in Backer and Backer, 2001a) could either stimulate tumor growth via VEGF

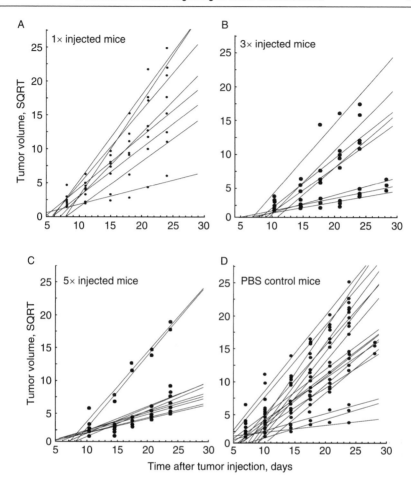

Figure 1.3. Inhibition of mouse mammary 4T1 tumors in mice given different dosing schedules of SLT–VEGF. 4T1 mouse breast carcinoma cells were implanted s.c. in the middle/lower backs of Balb/c mice. SLT–VEGF s.c. injections were given between the ears at 3.5–4.5 cm distances from growing tumors. A regression slope of 0.5 was chosen to classify mice in each treatment group as nonresponders (slope > 0.5) or responders (slope < 0.5). By this classification, 1× injected mice (A, single s.c. injection of 1×1 μg SLT–VEGF) have 1/8 responders, 3× injected mice (B, 3× 1 μg SLT–VEGF) have 4/9 responders, and 5× injections (C, 3× 1 μg followed by 2×0.33 μg SLT–VEGF) have 8/10 responders. The PBS group (D) has 3/23 responders. Comparing the proportion of responders versus nonresponders in the PBS control group against each SLT–VEGF treatment group by Fisher's exact test, the 1× injected group (p = 0.718) and 3× injected groups (p = 0.07) were not significantly different from the control group; whereas the 5× injected group (p = 0.0001) had a significantly higher proportion of responders.

activity, or inhibit tumor growth through some SLT-unrelated mechanisms. We found that multiple s.c. injections of 0.05 mg/kg/injection (1 µg/mouse/injection) of SLT–VEGFci did not affect 4T1 tumor growth, demonstrating the absolute requirement for SLT-based toxicity for tumor growth inhibition and lack of growth-promoting effects by the VEGF moiety (data not shown).

5. Systemic SLT–VEGF treatment targets endothelial cells with high levels of VEGFR-2 expression

To characterize the changes induced by SLT–VEGF in tumor vasculature, we performed a comparative immunohistochemical analysis of VEGFR-2 and a pan-endothelial marker CD31 in 4T1 and PC3 tumors. Immunofluorescent staining with a biotinylated tyramide amplification technique (as described by Backer et al., 2007) was used to enhance the detection sensitivity for both markers. We found that at these levels of sensitivity, VEGFR-2 was detected almost as readily as CD31. In view of reported detection of nonendothelial VEGFR-2 positive cells in some tumors (Jackson et al., 2002), we determined the degree of VEGFR-2 colocalization with CD31. This analysis revealed that in both tumors 75–80% of VEGFR-2 immunoreactivity is colocalized with CD31 (Fig. 1.4A). Furthermore, the colocalization reached ~89% and ~99% in the areas with the highest intensity of VEGFR-2 immunostaining (within top 20%, Fig. 1.4A), indicating that in both tumors high levels of VEGFR-2 were predominantly expressed by endothelial cells.

To selectively detect endothelial cells expressing the highest levels of VEGFR-2, we employed less sensitive chromogenic immunostaining. We found that chromogenic VEGR-2 immunostaining occupied a significantly smaller area than the fluorescent immunostaining. Only 10% of VEGFR-2 detectable by fluorescence immunohistostaining could be visualized with chromogenic development, indicating that this technique detected only cells with the highest VEGFR-2 expression levels (Fig. 1.4B). In contrast, chromogenic development of CD31 detected as much as 50% of CD31 immunoreactivity (Fig. 1.4B), providing information on a much larger portion of tumor vasculature than VEGFR-2 immunostaining.

Using chromogenic staining, we then compared responses to systemic SLT–VEGF treatment of a small subset of endothelial cells expressing the highest levels of VEGFR-2 and all endothelial cells. For this analysis, we have selected SLT–VEGF responders with tumors that were 2–2.5-fold smaller than tumors from control mice. Tumors were harvested 3–4 days after the last SLT–VEGF injection and a chromogenic immunostaining of VEGFR-2 and CD31 was performed on multiple alternating sections covering the bulk of selected tumors. To advance quantitative analysis from few microscopic fields to the bulk of

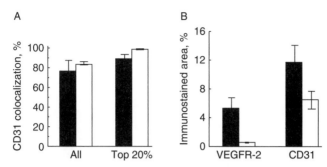

Figure 1.4. Immunohistochemical analysis of VEGFR-2 and CD31 expression in 4T1 and PC3 tumors. Tissue sections were probed with CD31 (PECAM) or VEGFR-2 (KDR/Flk-1) rat antimouse antibodies (Pharmingen, San Diego, CA). For chromogenic detection, secondary antibodies were coupled with peroxidase using the Elite streptavidin–biotin complex kit (Vector Labs) and visualized with DAB (Vector Labs). For fluorescent detection, visualization of secondary antibody was done by biotinylated tyramide amplification (TSA/HRP-streptavidin Kits, Molecular Probes). (A) Quantitative image analysis of VEGFR-2 and CD31 colocalization in tumor cryosections of 4T1 (closed) and PC3 (open) tumors. *Columns,* the percentage of VEGFR-2 immunostaining that colocalizes with CD31 immunostaining. *Bars,* ±S.D. of averaged values for five random fields. *All,* colocalization for all VEGFR-2 staining; *Top 20%,* colocalization for VEGFR-2 staining with the highest 20% intensity. (B) Quantitative image analysis of fluorigenic (closed) and chromogenic (open) VEGFR-2 and CD31 immunostaining of 4T1 tumor cryosections. *Columns,* the percentage of microscope field area occupied by immunostaining; *bars,* ±S.D. of averaged area values for five microscope fields.

tumor, immunostained sections were scanned using a flatbed scanner (Yang *et al.*, 1998) and digitized images were used for calculation of the total immunostained area as a percentage of the analyzed sectional area. Despite the large intrasectional variation (18–55%), in two independent experiments, SLT–VEGF treatment (five consecutive s.c. injections of 0.05 mg/kg/injection) decreased VEGFR-2 chromogenic immunostaining 4.9- ($p < 0.0003$) and 3.9-fold ($p < 0.0001$) (Fig. 1.5A). In contrast, analysis of the serial sections from the same tumors revealed that SLT–VEGF treatment either did not change ($p < 0.94$) or caused only a 1.4-fold ($p < 0.003$) decrease of CD31 immunostaining (Fig. 1.5B). Taken together, these data indicate that SLT–VEGF selectively depleted cells with high levels of VEGFR-2 expression from tumor vasculature, while the majority of CD31-positive cells remained unaffected. This observation was confirmed by a significant decrease of specific tumor accumulation of VEGF-based fluorescent tracer in 4T1 tumor-bearing mice treated with SLT–VEGF (Backer *et al.*, 2005).

If elimination of endothelial cells with high VEGFR-2 expression is a pharmacodynamic endpoint for SLT–VEGF action, it appears to be achieved fairly rapidly. In our *in vivo* experiments with PC3 and 4T1 tumors, three to five biweekly

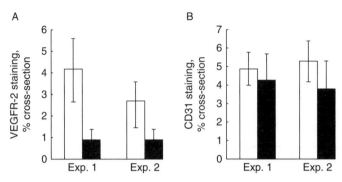

Figure 1.5. Systemic SLT–VEGF treatment selectively eliminates VEGFR-2 overexpressing tumor endothelial cells. In two independent experiments, mice injected with 0.5 million 4T1 cells/mouse received five s.c. injections of 1 μg SLT–VEGF or PBS on days 6, 8, 10, 12, 14 (Exp. 1), or 6, 8, 11, 13, 15 (Exp. 2), and sacrificed on day 18. In each experiment, a tumor from a control mouse and a 2–2.5-fold smaller tumor from a treated mouse were removed, cut vertically and sagittally, and two fragments with different orientations were snap-frozen. Frozen sections of 7-μm thickness were cut serially through ∼450 μm of the vertical and sagittal tumor fragments. Alternative sections were stained chromogenically for either CD31 or VEGFR-2. Slides were scanned with an HP Scanjet 5470c (Hewlett-Packard, Stamford, CT) flatbed scanner set to 2400 ppi hardware resolution. The number of pixels in immunostained areas was determined as a percentage of the total number of pixels in 75–80% of the whole cross section. Data are presented as average (±S.D.). Open columns, control PBS-treated mice. Closed columns, SLT–VEGF treated mice. (A) SLT–VEGF treatment dramatically decreases VEGFR-2 expression. Exp. 1, the average for vertical cross sections ($n=8$ for control, $n=5$ for treated mice) distributed over ∼115 μm. Exp. 2, the average for vertical cross sections ($n=8$, each group) distributed over ∼115 μm of tumor tissue. (B) SLT–VEGF treatment does not affect CD31 expression. Exp. 1, the average for vertical ($n=15$) and 15 sagittal ($n=15$) cross sections distributed over ∼450 μm. Exp. 2, the average for vertical ($n=8$) and sagittal ($n=8$) cross sections distributed over ∼450 μm of tumor tissue.

injections were enough to reach this effect. Importantly, SLT–VEGF-induced depletion of such cells was not rapidly reversible, as judged by immunohistochemical analysis undertaken 3–4 days after the last SLT–VEGF injection (data not shown).

6. Treatment with sunitinib and SLT–VEGF

Hypothetically, SLT–VEGF-induced selective depletion of VEGFR-2 overexpressing cells might be employed for inhibition of tumor angiogenesis, while remaining tumor vasculature can still be used for delivery chemotherapeutic drugs. On the other hand, SLT–VEGF might be used complementary to other drugs targeting VEGF/VEGFR-2 signaling pathway, to limit the resurgence of angiogenesis during "rest" periods or after prolonged treatment. For initial testing of the utility of combining SLT–VEGF with TKI inhibitors, we selected transgenic VEGFR-2 luc Balb/c nu/nu mice as hosts for human U87-MG glioma tumors. Mice were inoculated s.c. in the right flank with 3 million U87 MG cells. When tumors had reached a diameter of at least 5 mm, treatment the TKI inhibitor sunitinib was initiated. Mice were treated by gavage daily with 80 mg/kg sunitinib for 5 days and then treatment was discontinued and mice were allowed to recover for 3 days. Half of the mice received 5 μg doses of SLT–VEGF s.c. on days 3 and 6 after discontinuation of suntinib treatment and the other half received only saline. Immunohistochemical analysis of tumors harvested from mice in four experimental groups, control, sunitinib-treated, postsunitinib recovery, and postsunitinib recovery with SLT–VEGF treatment revealed a dramatic decline in CD31 and VEGFR-2 staining at the time of the last sunitinib injection. Staining of both markers after a 9-day recovery (Fig. 1.6, sunitinib/rest group) was significantly increased, approaching in the most vascularized areas that in untreated control mice. In contrast, in tumors from animals that underwent recovery coupled with SLT–VEGF treatment, the prevalence of CD31 and VEGFR-2 staining in general, and specifically in the highly vascularized areas, was significantly lower, reflecting inhibition of revascularization by SLT–VEGF (Fig. 1.6, sunitinib/SLT–VEGF group).

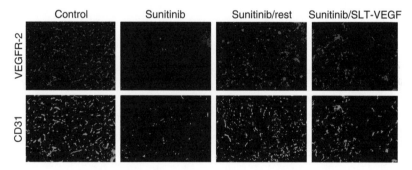

Figure 1.6. Consecutive treatment of tumor-bearing mice with sunitinib and SLT–VEGF. Transgenic VEGFR-2 luc Balb/c nu/nu mice were inoculated s.c. in the right flank with 3 million U87 MG human glioma cells. Treatment with sunitinib (80 mg/kg/daily, orally) was initiated when tumors had reached a diameter of at least 5 mm. After 5 days, treatment was discontinued, mice were allowed to rest for 3 days, and then treated with either SLT–VEGF (5 μg doses s.c.) on days 3 and 6, or saline. Mice from both groups were sacrificed on day 9 after the last sunitinib treatment and tumor cryosections were stained for immunofluorescent detection of CD31 (green) and VEGFR-2 (red) as described in the legend to Fig. 1.4, with fluorigenic development. Representative staining results within each group are shown. (See Page 1 in Color Section at the back of the book.)

V. SUMMARY

So far, the most important lesson taken from the experience with drugs targeting VEGF/VEGFR signaling pathway(s) is that in the clinic, as well as in the majority of experimental systems, these drugs only inhibit tumor growth, rather than lead to tumor regression. The most likely reason is that tumor endothelial cells can adapt to treatment and form functional vasculature. It is interesting that in this respect, tumor endothelial cells become, at least superficially, similar to normal endothelial cells that are not sensitive to inhibitors of VEGF/VEGFR signaling. It is tempting to speculate that only tumor cells, but not normal endothelial cells, are "addicted" to VEGF/VEGFR signaling, in the same sense as was proposed for tumor cells "addiction" to oncogene-controlled signaling (Weinstein and Joe, 2006, 2008). For such addicted cells, blocking only one pathway might lead to cell death, before other pathways are activated. In this paradigm, the "new" tumor endothelial cells that arise after prolonged treatment or after recovery period, do not acquire such addiction and therefore can function despite inhibition of VEGF/VEGFR signaling. However, as long as such cells express sufficient levels of VEGFR-2, they might remain sensitive to VEGF–toxin constructs that are internalized via VEGFR-2-mediated endocytosis. Indeed our preliminary experiments suggest that postsunitinib recovery of tumor vasculature can be inhibited by SLT–VEGF. Experiments are in progress to further explore the feasibility and efficacy of regimens that combine inhibitors of VEGF/VEGFR signaling and SLT–VEGF.

References

Al-Jaufy, A. Y., Haddad, J. E., King, S. R., McPhee, R. A., and Jackson, M. P. (1994). Cytotoxicity of a Shiga toxin A subunit-CD4 fusion protein to human immunodeficiency virus-infected cells. *Infect. Immun.* **62,** 956–960.

Al-Jaufy, A. Y., King, S. R., and Jackson, M. P. (1995). Purification and characterization of a Shiga toxin A subunit-CD4 fusion protein cytotoxic to human immunodeficiency virus-infected cells. *Infect. Immun.* **63,** 3073–3078.

Arora, N., Maood, R., Zheng, T., Cai, J., Smith, L., and Gill, P. S. (1999). Vascular endothelial growth factor chimeric toxin is highly active against endothelial cells. *Cancer Res.* **59,** 183–188.

Backer, M. V., and Backer, J. M. (2001a). Functionally active VEGF fusion proteins. *Protein Expr. Purif.* **23,** 1–7.

Backer, M. V., and Backer, J. M. (2001b). Targeting endothelial cells overexpressing VEGFR-2: Selective toxicity of Shiga-like toxin-VEGF fusion proteins. *Bioconj. Chem.* **12,** 1066–1073.

Backer, M. V., Budker, V. G., and Backer, J. M. (2001). Shiga-like toxin-VEGF fusion proteins are selectively cytotoxic to endothelial cells overexpressing VEGFR-2. *J. Control. Release* **74,** 349–355.

Backer, M. V., Gaynutdinov, T. I., Gorshkova, I. I., Crouch, R. J., Hu, T., Aloise, R., Arab, M., Przekop, K., and Backer, J. M. (2003). Humanized docking system for assembly of targeting drug delivery complexes. *J. Control. Release* **89,** 499–511.

Backer, M. V., Gaynutdinov, T. I., Patel, V., Bandyopadhyaya, A. K., Thirumamagal, B. T. S., Tjarks, W., Barth, R., Claffey, K. P., and Backer, J. M. (2005). Vascular endothelial growth factor selectively targets boronated dendrimers to tumor vasculature. *Mol. Cancer Ther.* **4**(9), 1423–1429.

Backer, M. V., Levashova, Z., Patel, V., Jehning, B. T., Claffey, K., Blankenberg, F. G., and Backer, J. M. (2007). Molecular imaging of VEGF receptors in angiogenic vasculature with single-chain VEGF based probes. *Nat. Med.* **13**, 504–509.

Bergers, G., and Hanahan, D. (2008). Modes of resistance to antiangiogenic therapy. *Nat. Rev. Cancer* **8**, 592–603.

Bikfalvi, A., Sauzeau, C., Moukadiri, H., Maclouf, J., Busso, N., Bryckaert, M., Plouet, J., and Tobelem, G. (1991). Interaction of vasculotropin/vascular endothelial cell growth factor with human umbilical vein endothelial cells: Binding, internalization, degradation, and biological effects. *J. Cell Physiol.* **149**, 50–59.

Blankenberg, F. G., Mandl, S., Cao, Y.-A., *et al.* (2004). Tumor imaging using a standardized radiolabeled adapter protein docked to vascular endothelial growth factor (VEGF). *J. Nucl. Med.* **45**, 1373–1380.

Blankenberg, F. G., Levashova, L., Sarkar, S. K., Pizzonia, J., Backer, M. V., and Backer, J. M. (2010). Noninvasive assessment of tumor VEGF receptors in response to treatment with pazopanib: A molecular imaging study. *Transl. Oncol.* **3**(1), in press.

Bozec, A., Gros, F. X., Penault-Llorca, F., Formento, P., Cayre, A., Dental, C., Etienne-Grimaldi, M. C., Fischel, J. L., and Milano, G. (2008). Vertical VEGF targeting: A combination of ligand blockade with receptor tyrosine kinase inhibition. *Eur. J. Cancer* **44**, 1922–1930.

Brigotti, M., Carnicelli, D., Alvergna, P., Mazzaracchio, R., Sperti, S., and Montanaro, L. (1997). The RNA-N-glycosidase activity of Shiga-like toxin I: Kinetic parameters of the native and activated toxin. *Toxiconolgy* **35**, 1431–1437.

Brown, L. F., Berse, B., Jackman, R. W., *et al.* (1995). Expression of vascular permeability factor (vascular endothelial growth factor) and its receptors in breast cancer. *Hum. Pathol.* **26**, 86–91.

Cai, W., Chen, K., Mohamedali, K. A., Cao, Q., Gambhir, S. S., Rosenblum, M. G., and Chen, X. (2006). PET of vascular endothelial growth factor receptor expression. *J. Nucl. Med.* **47**, 2048–2056.

Chang, Y. S., Adnane, J., Trail, P. A., Levy, J., Henderson, A., Xue, D., Bortolon, E., Ichetovkin, M., Chen, C., McNabola, A., Wilkie, D., Carter, C. A., *et al.* (2007). Sorafenib (BAY 43-9006) inhibits tumor growth and vascularization and induces tumor apoptosis and hypoxia in RCC xenograft models. *Cancer Chemother. Pharm.* **59**, 561–574.

Chu, Q. S. (2009). Aflibercept (AVE0005): An alternative strategy for inhibiting tumour angiogenesis by vascular endothelial growth factors. *Expert Opin. Biol. Ther.* **9**(2), 263–271.

Claffey, K. P., and Robinson, G. S. (1996). Regulation of VEGF/VPF expression in tumor cells: Consequences for tumor growth and metastasis. *Cancer Metastasis Rev.* **15**(2), 165–176.

Cohen, K. A., Liu, T., Bissonette, R., Puri, R. K., and Frankel, A. E. (2003). DAB389EGF fusion protein therapy of refractory glioblastoma multiforme. *Curr. Pharm. Biotechnol.* **4**, 39–49.

Couffinhal, T., Kearney, M., Witzenbichler, B., Chen, D., Murohara, T., Losordo, D. W., Symes, J., and Isner, J. M. (1997). Vascular endothelial growth factor/vascular permeability factor (VEGF/VPF) in normal and artherosclerotic human arteries. *Am. J. Pathol.* **150**, 1673–1685.

Dempke, W. C., and Heinemann, V. (2009). Resistance to EGF-R (erbB-1) and VEGF-R modulating agents. *Eur. J. Cancer* **45**(7), 1117–1128.

Deng, Q., and Barbieri, J. T. (2008). Molecular mechanisms of the cytotoxicity of ADP-ribosylating toxins. *Annu. Rev. Microbiol.* **62**, 271–288.

Duttaroy, A., Bourbeau, D., Wang, X. L., and Wang, E. (1998). Apoptosis rate can be accelerated or decelerated by overexpression or reduction of elongation factor-1α. *Exp. Cell Res.* **238**(1), 168–176.

Ebos, J. M., Lee, C. R., Cruz-Munoz, W., Bjarnason, G. A., Christensen, J. G., and Kerbel, R. S. (2009). Accelerated metastasis after short-term treatment with a potent inhibitor of tumor angiogenesis. *Cancer Cell.* **15**, 232–239.

Ellis, L. M., and Hicklin, D. J. (2008). VEGF-targeted therapy: Mechanisms of anti-tumour activity. *Nat. Rev. Cancer* **8**, 579–591.

Ferrara, N. (2009). Vascular endothelial growth factor. *Arterioscler. Thromb. Vasc. Biol.* 29(6), 789–791.

Folkman, J. (1995). Angiogenesis in cancer, vascular, rheumatoid and other disease. *Nat. Med.* **1**, 27–31.

Foss, F. M., Saleh, M. N., Krueger, J. G., Nichols, J. C., and Murphy, J. R. (1998). Diphtheria toxin fusion proteins. In "Clinical Applications of Immunotoxins" (A. E. Frankel, ed.), pp. 63–81. Springer, Berlin, Germany.

Frankel, A. E., Neville, D. M., Bugge, T. A., Kreitman, R. J., and Leppla, S. H. (2003). Immunotoxin therapy of hematologic malignancies. *Semin. Oncol.* **30**, 545–557.

Gourley, M., and Williamson, J. S. (2000). Angiogenesis: New targets for the development of anticancer chemotherapies. *Curr. Pharm. Des.* **6**, 417–439.

Grunwald, V., Soltau, J., Ivanyi, P., Rentschler, J., Reuter, C., and Drevs, J. (2009). Molecular targeted therapies for solid tumors: Management of side effects. *Onkologie* 32(3), 129–138.

Gurevich, F., and Perazella, M. A. (2009). Renal effects of anti-angiogenesis therapy: Update for the internist. *Am. J. Med.* 122(4), 322–328.

Hanahan, D., and Folkman, J. (1996). Patterns and emerging mechanisms of the angiogenic switch during tumorigenesis. *Cell* 86(3), 353–364.

Hayden, E. C. (2009). Cutting off cancer's supply lines. *Nature* **458**, 686.

Heckman, C. A., Holopainen, T., Wirzenius, M., Keskitalo, S., Jeltsch, M., Yla-Herttuala, S., Wedge, S. R., Jurgensmeier, J. M., and Alitalo, K. (2008). The tyrosine kinase inhibitor cediranib blocks ligand-induced vascular endothelial growth factor receptor-3 activity and lymphangiogenesis. *Cancer Res.* 68(12), 4754–4762.

Holzman, C. (2009). Whatever happened to immunotoxins? Research, and hope are still alive. *J. Natl. Cancer Inst.* 101(9), 624–625.

Hotz, H. G., Hotz, B., Bhargava, H., and Buhr, J. (2006). Specific targeting of tumor endothelial cells by a Shiga-like Toxin-VEGF fusion protein as a novel treatment strategy for pancreatic cancer. In "Deutsche Gesellschaft für Chirurgie" Vol. 35, pp. 5–6. Springer, Berlin, Heidelberg.

Hsu, A. R., Cai, W., Veeravagu, A., Mohamedali, K. A., Chen, K., Kim, S., Vogel, H., Hou, L. C., Tse, V., Rosenblum, M. G., and Chen, X. (2007). Multimodality molecular imaging of glioblastoma growth inhibition with vasculature-targeting fusion toxin VEGF121/rGel. *J. Nucl. Med.* 48(3), 445–454.

Iordanov, M. S., and Magun, B. E. (1999). Different mechanisms of c-Jun NH(2)-terminal kinase-1 (JNK1) activation by ultraviolet-B radiation and by oxidative stressors. *J. Biol. Chem.* **274**, 25801–25806.

Iordanov, M. S., Pribnow, D., Magun, J. L., Dinh, T. H., Pearson, J. A., Chen, S. L., and Magun, B. E. (1997). Ribotoxic stress response: Activation of the stress-activated protein kinase JNK1 by inhibitors of the peptidyl transferase reaction and by sequence-specific RNA. *Mol. Cell. Biol.* **17**, 3373–3381.

Izzedine, H., Ederhy, S., Goldwasser, F., Soria, J. C., Milano, G., Cohen, A., Khayat, D., and Spano, J. P. (2009). Management of hypertension in angiogenesis inhibitor-treated patients. *Ann. Oncol.* 20(5), 807–815.

Jackson, M. W., Roberts, J. S., Heckford, S. E., *et al.* (2002). A potential autocrine role for vascular endothelial growth factor in prostate cancer. *Cancer Res.* **62**, 854–859.

Jain, R. K. (2005). Normalization of tumor vasculature: An emerging concept in antiangiogenic therapy. *Science* **307**, 58–62.

Jain, R. K., Duda, D. G., Clark, J. W., and Loeffler, J. S. (2006). Lessons from phase III clinical trials on anti-VEGF therapy for cancer. *Nat. Clin. Pract. Oncol.* **1**, 24–40.

Johannsen, M., Florcken, A., Bex, A., Roigas, J., Cosentino, M., Ficarra, V., Kloeters, C., Rief, M., Rogalla, P., Miller, K., and Grunwald, V. (2009). Can tyrosine kinase inhibitors be discontinued in patients with metastatic renal cell carcinoma and a complete response to treatment? A multi-centre, retrospective analysis. *Eur. Urol.* **55**, 1430–1439.

Jubb, A. M., Oates, A. J., Holden, S., and Koeppen, H. (2006). Predicting benefit from antiangiogenic agents in malignancy. *Nat. Rev. Cancer* **6**, 626–635.

Kaplan, B. S., Cleary, T. G., and Obrig, T. G. (1990). Recent advances in understanding the pathogenesis of the hemolytic uremic syndromes. *Pediatr. Nephrol.* **4**, 276–283.

Kerbel, R. S., and Kamen, B. A. (2004). The anti-angiogenesis basis of metronomic chemotherapy. *Nat. Rev. Cancer* **4**, 425–436.

Koukourakis, M. I., Giatromanolaki, A., Thorpe, P. E., et al. (2000). Vascular endothelial growth factor/KDR activated microvessel density versus CD31 standard microvessel density in non-small cell lung cancer. *Cancer Res.* **60**, 3088–3095.

Kreitman, R. J. (2006). Immunotoxins for targeted cancer therapy. *AAPS J.* **8**(3), E532–E551.

Kreitman, R. J. (2009). Recombinant immunotoxins containing truncated bacterial toxins for the treatment of hematologic malignancies. *BioDrugs* **23**(1), 1–13.

Kumar, R., Knick, V. B., Rudolph, S. K., Johnson, J. H., Crosby, R. M., Crouthamel, M.-C., Hopper, T. M., Miller, C. G., Harrington, L. E., Onori, J. A., Mullin, R. J., Gilmer, T. M., et al. (2007). Pharmacokinetic-pharmacodynamic correlation from mouse to human with pazopanib, a multikinase angiogenesis inhibitor with potent antitumor and antiangiogenic activity. *Mol. Cancer Ther.* **6**, 2012–2021.

Levashova, Z., Backer, M., Backer, J. M., and Blankenberg, F. G. (2008). Direct labeling of Cys-tag in scVEGF with technetium 99m. *Bioconjug. Chem.* **19**, 1049–1054.

Lin, M. I., and Sessa, W. C. (2004). Antiangiogenic therapy: Creating a unique "window" of opportunity. *Cancer Cell* **6**, 529–531.

Liu, T. F., Cohen, K. A., Ramag, J. G., Willingham, M. C., Thorburn, A. M., and Frankel, A. E. (2003). A diphtheria toxin-epidermel growth factor fusion protein is cytotoxic to human glioblastoma multiforme cells. *Cancer Res.* **63**(8), 1834–1837.

Loges, S., Mazzone, M., Hohensinner, P., and Carmeliet, P. (2009). Silencing or fueling metastasis with VEGF inhibitors: Antiangiogenesis revisited. *Cancer Cell* **15**, 167–170.

Lohela, M., Bry, M., Tammela, T., and Alitalo, K. (2009). VEGFs and receptors involved in angiogenesis versus lymphangiogenesis. *Curr. Opin. Cell Biol.* **21**(2), 154–165.

Mancuso, M. R., Davis, R., Norberg, S. M., O'Brien, S., Sennino, B., Nakahara, T., Yao, V. J., Inai, T., Brooks, P., Freimark, B., Shalinsky, D. R., Hu-Lowe, D. D., et al. (2006). Rapid vascular regrowth in tumors after reversal of VEGF inhibition. *J. Clin. Invest.* **116**, 2610–2621.

Manley, P. W., Bold, G., Bruggen, J., et al. (2004). Advances in the structural biology, design and clinical development of VEGF-R kinase inhibitors for the treatment of angiogenesis. *Biochim. Biophys. Acta* **1697**, 17–27.

Mendel, D. B., Laird, A. D., Xin, X., Louie, S. G., Christensen, J. G., Li, G., Schreck, R. E., Abrams, T. J., Ngai, T. J., Lee, L. B., Murray, L. J., Carver, J., et al. (2003). In vivo antitumor activity of SU11248, a novel tyrosine kinase inhibitor targeting vascular endothelial growth factor and platelet-derived growth factor receptors: Determination of a pharmacokinetic/pharmacodynamic relationship. *Clin. Cancer Res.* **9**, 327–337.

Mohamedali, K. A., Poblenz, A. T., Sikes, C. R., Navone, N. M., Thorpe, P. E., Darnay, B. G., and Rosenblum, M. G. (2006). Inhibition of prostate tumor growth and bone remodeling by the vascular targeting agent VEGF121/rGel. *Cancer Res.* **66**(22), 10919–10928.

Obrig, T. G., Del Vecchio, P. J., Karmali, M. A., Petric, M., Moran, T. P., and Judge, T. K. (1987). Pathogenesis of haemolytic uraemic syndrome (Letter). *Lancet* **2**, 687.

Obrig, T., Louise, C. B., Lingwood, C. A., Boyd, B., Barley-Maloney, L., and Daniel, T. O. (1993). Endothelial heterogeneity in Shiga toxin receptors and responses. *J. Biol. Chem.* **268**, 15484–15488.

Olsnes, S., Reisbig, R., and Eiklid, K. (1981). Subunit structure of Shigella cytotoxin. *J. Biol. Chem.* **256,** 8732–8738.

Olson, T. A., Mohanraj, D., Roy, S., and Ramakrishnan, S. (1997). Targeting the tumor vasculature: Inhibition of tumor growth by a vascular endothelial growth factor-toxin conjugate. *Int. J. Cancer* **73,** 865–870.

Padera, T. P., Kuo, A. H., Hoshida, T., Liao, S., Lobo, J., Kozak, K. R., Fukumura, D., and Jain, R. K. (2008). Differential response of primary tumor versus lymphatic metastasis to VEGFR-2 and VEGFR-3 kinase inhibitors cediranib and vandetanib. *Mol. Cancer Ther.* **7**(8), 2272–2279.

Paez-Ribes, M., Allen, E., Hudock, J., Takeda, T., Okuyama, H., Vinals, F., Inoue, M., Bergers, G., Hanahan, D., and Casanovas, O. (2009). Antiangiogenic therapy elicits malignant progression of tumors to increased local invasion and distant metastasis. *Cancer Cell* **15,** 220–231.

Palmowski, M., Huppert, J., Hauff, P., Reinhardt, M., Schreiner, K., Socher, M. A., Hallscheidt, P., Kauffmann, G. W., Semmler, W., and Kiessling, F. (2008). Vessel fractions in tumor xenografts depicted by flow- or contrast-sensitive three-dimensional high-frequency Doppler ultrasound respond differently to antiangiogenic treatment. *Cancer Res.* **68,** 7042–7049.

Pastan, I., Hassan, R., Fitzgerald, D. J., and Kreitman, R. J. (2006). Immunotoxin therapy of cancer. *Nat. Rev. Cancer* **6**(7), 559–565.

Peifer, C., Buhler, S., Hauser, D., Kinkel, K., Totzke, F., Schachtele, C., and Laufer, S. (2009). Design, synthesis and characterization of N9/N7-substituted 6-aminopurines as VEGF-R and EGF-R inhibitors. *Eur. J. Med. Chem.* **44**(4), 1788–1793.

Pennell, N. A., and Lynch, T. J., Jr. (2009). Combined inhibition of the VEGFR and EGFR signaling pathways in the treatment of NSCLC. *Oncologist* **14**(4), 399–411.

Potala, S., Sahoo, S. K., and Verma, R. S. (2008). Targeted therapy of cancer using diphtheria toxin-derived immunotoxins. *Drug Discov. Today* **13**(17–18), 807–815.

Rafii, S., Lyden, D., Benezra, R., Hattori, K., and Heissig, B. (2002). Vascular and haematopoietic stem cells: Novel targets for anti-angiogenesis therapy? *Nat. Rev. Cancer* **2,** 826–835.

Ramakrishnan, S., Olson, T. A., Bauch, V. L., and Mohanraj, D. (1996). Vascular endothelial growth factor-toxin conjugate specifically inhibits KDR/flk-1 positive endothelial cell proliferation *in vitro* and angiogenesis *in vivo. Cancer Res.* **56,** 1324–1330.

Ran, S., Mohamedali, K. A., Luster, T. A., Thorpe, P. E., and Rosenblum, M. G. (2005). The vascular-ablative agent VEGF(121)/rGel inhibits pulmonary metastases of MDA-MB-231 breast tumors. *Neoplasia (New York)* **7**(5), 486–496.

Richardson, S. E., Karmali, M. A., Becker, L. E., and Smith, C. R. (1988). *Hum. Pathol.* **19,** 1102–1108.

Ruan, J., Hajjar, K., Rafii, S., and Leonard, J. P. (2009). Angiogenesis and antiangiogenic therapy in non-Hodgkin's lymphoma. *Ann. Oncol.* **20**(3), 413–424.

Saaristo, A., Karpanen, T., and Alitalo, K. (2000). Mechanisms of angiogenesis and their use in the inhibition of tumor growth and metastasis. *Oncogene* **19,** 6122–6129.

Sandvig, K., Olsnes, S., Brown, J., Peterson, O., and Van Beurs, B. (1989). *J. Cell Biol.* **108,** 1331–1343.

Saxena, S. K., O'Brien, A. D., and Ackerman, E. J. (1989). Shiga toxin, Shiga-like toxin II variant, and ricin are all single-site RNA N-glycosidases of 28 S RNA when microinjected into Xenopus oocytes. *J. Biol. Chem.* **264,** 596–601.

Schmidt, U., Ahmed, J., Michalsky, E., Hoepfner, M., and Preissner, R. (2008). Comparative VEGF receptor tyrosine kinase modeling for the development of highly specific inhibitors of tumor angiogenesis. *Genome Inf. Serv.* **20,** 243–251.

Siemann, D. W., Bibby, M. C., Dark, G. G., Dicker, A. P., Eskens, F. A., Horsman, M. R., Marme, D., and Lorusso, P. M. (2005). Differentiation and definition of vascular-targeted therapies. *Clin. Cancer Res.* **11**(2 Pt 1), 416–420.

Sledge, G. W., Jr, and Miller, K. D. (2002). Angiogenesis and antiangiogenic therapy. *Curr. Probl. Cancer* **26,** 1–60.

Sloan, B., and Scheinfeld, N. S. (2008). Pazopnib, a VEGF receptor tyrosine kinase inhibitor for cancer therapy. *Curr. Opin. Invest. Drugs* **9**, 1324–1335.

Smith, N. R., James, N. H., Oakley, I., Wainwright, A., Copley, C., Kendrew, J., Womersley, L. M., Jürgensmeier, J. M., Wedge, S. R., and Barry, S. T. (2007). Acute pharmacodynamic and antivascular effects of the vascular endothelial growth factor signaling inhibitor AZD2171 in Calu-6 human lung tumor xenografts. *Mol. Cancer Ther.* **6**(8), 2198–2208.

Srivastava, S. K., Kumar, V., Agarwal, S. K., Mukherjee, R., and Burman, A. C. (2009). Synthesis of quinazolines as tyrosine kinase inhibitors. *Curr. Med. Chem. AntiCancer Agents* **9**(3), 246–275.

Stirpe, F., and Battelli, M. G. (2006). Ribosome-inactivating proteins: Progress and problems. *Cell Mol. Life Sci.* **63**(16), 1850–1866.

Stirpe, F., Olsnes, S., and Pihl, A. (1980). Gelonin, a new inhibitor of protein synthesis, nontoxic to intact cells. Isolation, characterization, and preparation of cytotoxic complexes with concanavalin A. *J. Biol. Chem.* **255**, 6947–6953.

Talpaz, M., Kantarjian, H., Freireich, E., Lopez, V., Zhang, W., Cortes-Franco, J., Scheinberg, D., and Rosenblum, M. G. (2003). Phase I clinical trial of the anti-CD-33 immunotoxin HuM195/rGel. *Am. Assoc. Cancer Res.* **44**, 1066.

Terman, B. I., and Dougher-Vermazen, M. (1996). Biological properties of VEGF/VPF receptors. *Cancer Metastasis Rev.* **15**(2), 159–163.

Theodoulou, M., Baselga, J., Scher, H., Dantis, L., Trainor, K., Mendelsohn, J., Howes, L., Elledge, R., Ravdin, P., Bacha, P., Brandt-Sarif, T., and Osborne, K. (1995). Phase I dose escalation study of the safety, tolerability, pharmacokinetics and biologic effects of DAB389-EGF in patient with solid malignances that express EGF receptors. *Proc. ASCO* **14**, 480.

Thorpe, P. E., Chaplin, D. J., and Blakey, D. C. (2003). The first international converence on vascular targeting: Meeting overview. *Cancer Res.* **63**(5), 1144–1147.

Vallera, D. A., Panoskaltsis-Mortari, A., and Blazar, B. R. (1997). Renal dysfunction accounts for the dose limiting toxicity of DT390anti-CD3sFv, a potential new recombinant anti-GVHD immunotoxin. *Protein Eng.* **10**, 1071–1076.

van Cruijsen, H., van der Veldt, A., and Hoekman, K. (2009). Tyrosine kinase inhibitors of VEGF receptors: Clinical issues and remaining questions. *Front. Biosci.* **14**, 2248–2268.

Veenendaal, L. M., Jin, H., Ran, S., *et al.* (2002). *In vitro* and *in vivo* studies of a VEGF121/rGelonin chimeric fusion toxin targeting the neovasculature of solid tumors. *Proc. Natl. Acad. Sci. USA* **99**, 7866–7871.

Veikkola, T., and Alitalo, K. (1999). VEGFs, receptors and angiogenesis. *Semin. Cancer Biol.* **9**(3), 211–220.

Veikkola, T., Karkkainen, M., Claesson-Welsh, L., and Alitalo, K. (2000). Regulation of angiogenesis via vascular endothelial growth factor receptors. *Cancer Res.* **60**, 203–212.

Wang, H., Cai, W., Chen, K., Li, Z.-B., Kashefi, A., He, L., and Chen, X. (2007). A new PET tracer specific for vascular endothelial growth factor receptor 2. *Eur. J. Nucl. Med. Mol. Imaging* **34**, 2001–2010.

Wang, H., Chen, K., Niu, G., and Chen, X. (2009). Site-specifically biotinylated VEGF(121) for near-infrared fluorescence imaging of tumor angiogenesis. *Mol. Pharm.* **6**(1), 285–294.

Weinstein, I. B., and Joe, A. K. (2006). Mechanisms of disease: Oncogene addiction—A rationale for molecular targeting in cancer therapy. *Nat. Clin. Pract. Oncol.* **3**, 448–457.

Weinstein, I. B., and Joe, A. K. (2008). Oncogene addiction. *Cancer Res.* **68**, 3077–3080.

Williams, J. M., Lea, N., Lord, J. M., Roberts, L. M., Milford, D. V., and Taylor, C. M. (1997). Comparison of ribosome-inactivating proteins in the induction of apoptosis. *Toxicol. Lett.* **91**(2), 121–127.

Yang, Y., Shuaib, A., and Li, Q. (1998). Quantification of infarct size on focal cerebral ischemia model of rats using a simple and economical method. *J. Neurosci. Methods* **84**, 9–16.

2

Adeno-Associated Viral Vectors and Their Redirection to Cell-Type Specific Receptors

Stefan Michelfelder and Martin Trepel

Department of Oncology and Hematology, Hubertus Wald Cancer Center,
University Medical Center Hamburg-Eppendorf, Martinistrasse 52,
D-20246 Hamburg, Germany

Advances in Genetics, Vol. 67
Copyright 2009, Elsevier Inc. All rights reserved.

0065-2660/09 $35.00
DOI: 10.1016/S0065-2660(09)67002-4

ABSTRACT

Efficient and specific delivery of genes to the cell type of interest is a crucial issue in gene therapy. Adeno-associated virus (AAV) has gained particular interest as gene vector recently and is therefore the focus of this chapter. Its low frequency of random integration into the genome and the moderate immune response make AAV an attractive platform for vector design. Like in most other vector systems, the tropism of AAV vectors limits their utility for certain tissues especially upon systemic application. This may in part be circumvented by using AAV serotypes with an *in vivo* gene transduction pattern most closely fitting the needs of the application. Also, the tropism of AAV capsids may be changed by combining parts of the natural serotype diversity. In addition, peptides mediating binding to the cell type of interest can be identified by random phage display library screening and subsequently be introduced into an AAV capsid region critical for receptor binding. Such peptide insertions can abrogate the natural tropism of AAV capsids and result in detargeting from the liver *in vivo*. In a novel approach, cell type-directed vector capsids can be selected from random peptide libraries displayed on viral capsids or serotype-shuffling libraries *in vitro* and *in vivo* for optimized transduction of the cell type or tissue of interest. © 2009, Elsevier Inc.

I. INTRODUCTION

The biological safety and the unsatisfactory efficacy of gene transfer vectors pose significant challenges for gene therapy (Somia and Verma, 2000; Thomas *et al.*, 2003). Problems of current gene therapy vectors include unintended transduction of certain tissues, adverse immune reactions, and lack of efficient transduction of the cells of interest (Trepel *et al.*, 2000a). Targeting vectors to certain cell types or tissues may have a very high impact on all these parameters. Recombinant adeno-associated virus (AAV) vectors have become very popular as potential gene therapy vectors because of their ability to mediate stable and efficient gene expression, combined with a favorable biological safety profile. Like in other vector systems, however, their lack of specificity poses a serious problem that has been challenged during the last decade by a plethora of different technical approaches to change their tropism. These will be reviewed here and comprise the use of the various AAV serotypes, the insertion of targeting ligands into the AAV capsid, as well as the development of various kinds of combinatorial AAV capsid libraries for selection of targeted and transduction-optimized AAV vectors. All of these approaches have their unique advantages and disadvantages, and their specific use depends very much on the intended application the vectors are needed for.

A. AAV as virus and vector

AAV is a small, nonenveloped DNA virus which belongs to the parvovirus family. To date, 12 different serotypes have been isolated from primate or human tissues (Grimm and Kay, 2003; Schmidt et al., 2008; Wu et al., 2006a). From all serotypes described so far, AAV-2 is the one that has been best characterized. As a wild-type virus, AAV is nonpathogenic, only mildly immunogenic, and has the potential to integrate site-specifically into the host genome. Its broad host tropism allows for efficient gene delivery in various target tissues.

1. AAV genomic organization

AAV harbors a single-stranded genome of approximately 4.7 kb that comprises two open reading frames (*rep* and *cap*) flanked by inverted terminal repeats (ITRs). The palindromic nucleotide sequence of the ITRs forms a characteristic T-shaped hairpin structure serving as structural *cis*-acting elements that are required for viral genome replication, packaging, and rescue from the integrated state. In addition, the ITRs have regulatory influence on viral gene expression and host genome integration (Goncalves, 2005).

The nonstructural Rep proteins Rep78 and Rep52, and their respective splice variants Rep68 and Rep40 are under transcriptional control of two promoters (p5 and p19) and play an essential role in viral DNA replication and packaging, regulation of gene expression, and site-specific integration (Goncalves, 2005). The *cap* gene encodes for three structural capsid proteins VP1, VP2, and VP3 (90, 72, 62 kDa, respectively) that share the same C-terminal domain, while VP2 and VP1 contain additional N-terminal sequences. The N terminus of VP1 contains a phospholipase A2 (PLA2) domain necessary for endosomal escape and nuclear entry during the viral infection (Bleker et al., 2005; Girod et al., 2002; Sonntag et al., 2006). The N terminus of VP1, VP2, and VP3 contains four basic regions (BR) that constitute putative nuclear localization sequences (NLS) presumably involved in the nuclear transfer of the virus (Grieger et al., 2006, 2007; Vihinen-Ranta et al., 2002).

2. AAV structure

The AAV capsid is composed of 60 subunits of the VP proteins VP1, VP2, and VP3 at a molar ratio of approximately 1:1:10 and generates a $T = 1$ icosahedral capsid with approximately 25 nm in diameter. The atomic structure of various serotypes has been resolved by X-ray crystallography (AAV-1, -2, -4, and -8) (Govindasamy et al., 2006; Miller et al., 2006; Nam et al., 2007; Xie et al., 2002) or by cryoelectron microscopy (AAV-5) (Walters et al., 2004), and the crystal structures for AAV-6, -7, and -9 are currently being resolved (Mitchell et al., 2009; Quesada et al., 2007;

Xie *et al.*, 2008). All serotypes share a common capsid structure comprising a conserved eight-stranded antiparallel β-barrel (labeled B-I) motif with large loop insertions between the β-strands that constitute the basic structure of the capsid (Fig. 2.1). In intact AAV capsids, three capsid subunits contribute to the loops

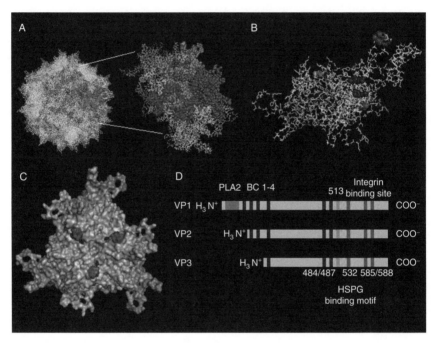

Figure 2.1. Structure of the AAV-2 capsid and amino acids involved in binding to cellular receptors. (A) Surface topology of the AAV capsid composed of 60 VP protein subunits. Three VP subunits contribute to the formation of characteristic threefold spikes clustering around the threefold axis of symmetry. Further, a cylindrical pore structure clustered around the fivefold axis of symmetry, surrounded by characteristic depressions termed as canyon, plateau, and dimple, form the structure of the AAV surface that interact with cellular receptors (image adopted from http://www.virology.wisc.edu). (B) Model of the VP-3 capsid protein and localization of the amino acids involved in receptor binding on the AAV-2 capsid surface. The basic amino acids R585/R588 (red), R484/R487 (blue), and K532 (green) form the heparin sulfate proteoglycan (HSPG) binding motif. The amino acids 511–513 (yellow) forming the NGR motif are proposed to serve as a binding site for integrin $\alpha 5\beta 1$. (C) Plan view on a VP trimer and a characteristic threefold spike region and surrounding regions. Localization of amino acids involved in binding to HSPG and integrin $\alpha 5\beta 1$ are color-coded as (B). (D) Schematic depiction of the functional domains of the three structural AAV proteins VP1–VP3. VP1 contains a phospholipase A2 (PLA2) domain, four basic regions (BR1–4) are located at the N-terminus of VP1–VP3. The HSPG binding domain is generated by the basic residues at positions R484, R487, K532, R585, and R588 which are located near the C-terminus of the VP proteins. The NGR motif 511–513 forms an integrin $\alpha 5\beta 1$ binding domain. (See Page 1 in Color Section at the back of the book.)

between the subunits G and H, leading to the formation of characteristic protrusions which are arranged in groups of three ("threefold spikes") and cluster around the threefold axis of symmetry. Another characteristic feature is the cylindrical pore structure which is clustered around the fivefold axis of symmetry and is surrounded by characteristic depressions termed as canyon, plateau, and dimple (Kern et al., 2003; Lochrie et al., 2006; Opie et al., 2003). Differences in amino acid sequences within the variable loop regions characterize the different AAV serotypes (ranging from ~ 1% to 45%) (Gao et al., 2002; Mori et al., 2004). These loop regions presumably mediate most if not all of the interactions of the AAV capsid with cellular receptors. In addition, they can serve as immune epitopes.

Mutagenesis-based approaches and structural data facilitated the identification of receptor-binding sites within the capsid. The amino acids involved in binding of AAV serotype 2 (AAV-2) to its primary cellular receptor heparan sulfate-proteoglycan (HSPG) are presented within two adjacent VP protein subunits forming protrusions on the threefold spike. This binding site comprises a cluster of five basic amino acids at amino acid positions R484, R487, K532, R585, and R588 (VP numbering) (Fig. 2.1). Positively charged amino acid residues bind negatively charged sulfate and carboxyl groups of HSPG mainly via electrostatic interactions (Kern et al., 2003; Opie et al., 2003; Xie et al., 2002). Of note, amino acids adjacent to the heparan sulfate binding site can play a crucial role for immune recognition by the host organism. Point mutations at position R471A or N587A or insertion of a small peptide ligand at the latter position enabled AAV capsids to escape binding and neutralizing by antibodies (Lochrie et al., 2006; Wobus et al., 2000). Further, the peptide string RXXR at positions 585–588 is involved in the cellular uptake into dendritic cells and the activation of capsid-specific T-cells, thereby contributing to the generation of host cytotoxic T-cell responses (Vandenberghe et al., 2006). The cryo-EM structure of AAV and the heparin complex suggest that the footprint that generates the HSPG binding motif is wider than apparent through mutagenesis-based approaches and extends into the "dead zone" region—a site that has been proposed to be a potential binding site for AAV co-receptors (Lochrie et al., 2006; O'Donnell et al., 2009). These observations make it difficult to confine particular capsid domains that interact with cellular co-receptors, since mutations in the HSPG binding region may also affect possible co-receptor interactions.

The amino acids NGR at positions 511–513 adjacent to the HSPG binding site is highly conserved in several serotypes (except for AAV serotypes -4, -5, and -11) and has been proposed to serve as a binding site for integrin $\alpha 5\beta 1$. (Asokan et al., 2006) (Fig. 2.1).

3. Cellular receptors

The cellular transduction by AAV is initiated by binding of the viral capsid to cell surface receptors. Various glycan motifs that are widely expressed on various cell types and tissues can be used as attachment receptor for different AAV serotypes. The primary receptor for AAV-2 and -3 is cellular HSPG (Summerford and Samulski, 1998), while serotypes AAV-1, -4, -5, and -6 use different derivates of O- or N-linked sialic acid, respectively (Kaludov et al., 2001; Seiler et al., 2006; Wu et al., 2006a,b). However, for AAV-2 and -3, cellular transduction of AAV has also been demonstrated in the absence of HSPG (Boyle et al., 2006; Handa et al., 2000). Furthermore, mutations of the AAV-2 capsid in the HSPG binding domain result in an heparin binding deficient AAV phenotype, leading to an increased transduction of the heart in vivo (Kern et al., 2003). Both findings suggested that other receptors than HSPG are involved in the viral attachment. To date, integrin $\alpha v\beta5$, integrin $\alpha5\beta1$, hepatocyte growth factor receptor (c-Met), and CD9 have been described as co-receptors for AAV-2 infection (Asokan et al., 2006; Kashiwakura et al., 2005; Kurzeder et al., 2007; Summerford et al., 1999), while the fibroblast growth factor receptor-1 (FGFR1) is utilized both by AAV-2 and -3 (Qing et al., 1999). The 37/67-kDa laminin receptor (LamR) can serve as co-receptor for AAV-2, -3, -8, and -9 (Akache et al., 2006a,b) and platelet derived growth factor receptor (PDGFR) mediates AAV-5 transduction (Di Pasquale et al., 2003).

4. Infection and replicative cycle

For cellular entry, AAV depends on the interaction with various co-receptors. It has been proposed that HSPG binding to AAV-2 leads to structural rearrangements of the viral capsid providing the conformational basis for interaction with integrin $\alpha5\beta1$ required for receptor-mediated endocytosis (Asokan et al., 2006). However, structural analysis of the AAV-2 heparin complex did not provide evidence for receptor-induced conformational changes (O'Donnell et al., 2009). Nevertheless, AAV-2 interaction to cellular co-receptors initiates cell entry of the virions via clathrin-coated pits in a dynamin-dependent process into endosomes (Bartlett et al., 2000; Ding et al., 2005). AAV binding to cell surface receptors is required to activate the phosphatidylinositol-3-kinase pathway via Rac1 that triggers the intracellular trafficking of AAV to the nucleus along microtubules and microfilaments (Sanlioglu et al., 2000).

      Vesicular trafficking to the nuclear area is a slow and rate-limiting step in AAV gene transduction in some if not most cell types (Duan et al., 2000; Hansen et al., 2000, 2001). Six potential pathways have been described for AAV-2 and -5. Following viral internalization, virions are taken up into early endosomes and can move either to late endosomal compartments where AAV accumulates in early lysosomes or the *trans*-Golgi network. Depending on the cell type and the viral dose, AAV can also accumulate in perinuclear recycling endosomes (PNRE) or traveling from PNRE to the *trans*-Golgi network, or even exit from very early endosomes (Ding *et al.*, 2005). Endosomal proteases like Cathepsin B and L have significant influence on AAV gene transfer efficiency, possibly through selective cleavage of capsid proteins for further uncoating steps (Akache *et al.*, 2006b). As AAV-5 internalization can also occur via the caveolar pathway, virions have been detectable in caveosomes (Bantel-Schaal *et al.*, 2009).

Endosomal release of the virions to the cytoplasm occurs when conformational changes of the VP1/VP2 N-termini leads to the exposure of a phospholipase A2 (PLA2) domain within the capsid through pores located at the fivefold axis of symmetry. Acidification of the endosomal compartment seems likely to trigger this process but is not sufficient for the conformational change of VP1/VP2, suggesting that further, so far unknown, mediators are involved (Kronenberg *et al.*, 2005; Sonntag *et al.*, 2006). Interestingly, sequence analyses of the N-terminal amino acids of VP1 among different serotypes may suggest that the phopholipase A2 is a current feature of AAV (Ding *et al.*, 2005). After endosomal escape and release into the cytosol, AAV accumulates in a perinuclear pattern before its translocation into the nucleus occurs. During this step of post-entry processing, ubiquitylation and ubiquitin-dependent degradation of the viral capsid by proteasomes can influence the level of gene transduction in different cell types. Consequently, proteasome inhibitors have been used to increase the rate of AAV-2 transduction in certain cell types and tissues (Duan *et al.*, 2000; Grieger and Samulski, 2005; Jennings *et al.*, 2005; Yan *et al.*, 2002). Mutations of surface-exposed tyrosine phosphorylation sites can lead to a protection from proteasomal-mediated degradation of the capsid thereby increasing transduction efficiency of AAV-2 (Zhong *et al.*, 2008). While it has been shown that the cytoplasmic movement of AAV is mediated via cytoskeletal ATP-dependent motor proteins and/or Brownian diffusion (Ding *et al.*, 2005), the mechanism of nuclear translocation remains unclear. Possibly, the N terminus of VP1/VP2 serves as putative nuclear translocation signal that mediates viral entry via the nuclear pore complex (NPC) (Grieger *et al.*, 2006). However, alternative pathways for nuclear translocation are also under discussion (Hansen *et al.*, 2001; Xiao *et al.*, 2002). The presence of intact virions in the nucleus suggests that uncoating of AAV occurs in the nucleus. It is likely that mobilization of intact AAV capsids from the nucleolus to nucleoplasmic sites permits uncoating and might determine the ratio of single-stranded genomes that become double stranded (Johnson and Samulski, 2009). Depending on the host cell DNA synthesis machinery, second strand synthesis is another major rate-limiting step in AAV gene transduction. The conversion of single-stranded AAV genome to double-stranded DNA is required for viral gene expression, stabilization of the genome, and prevention of degradation. When latent,

AAV-2 persists either by Rep protein-mediated site-specific integration into the q-arm of chromosome 19 (AAVS1), a region adjacent to muscle-specific genes or as circular extrachromosal episomes (Kotin et al., 1992; Samulski et al., 1991; Schnepp et al., 2005). For productive replication, AAV requires helper viral proteins delivered by adenovirus (Ad) or herpes simplex virus (HSV) that enables the rescue of the AAV genome, DNA replication, and gene expression of the viral proteins. Capsid assembly takes place in the nucleoli of infected cells that are finally redistributed to the nucleoplasm (Hunter and Samulski, 1992; Wistuba et al., 1997). There, virions are colocalized with Rep 78/68-tagged viral DNA. Rep 52/40 proteins are involved in unwinding and transfer of the viral DNA into the empty capsid through pores located at the fivefold axes of symmetry (Bleker et al., 2006; King et al., 2001). Finally, replicated viruses are released by lysis of the host cell.

5. AAV-derived gene vectors

Recombinant AAV (rAAV) vectors are constructed by replacement of the viral DNA containing the two open reading frames rep and cap with an expression cassette encoding the gene of interest under transcriptional control of a suitable promoter. The ITR sequences required for replication and packaging are the sole remainder of the wild-type virus. For vector production, the structural and nonstructural Rep and Cap proteins must be provided in trans. Today, vectors are usually produced by transfection of a suitable cell line with three vector plasmids: (i) the expression cassette flanked by the ITRs, (ii) the rep cap helper sequences, and (iii) the adenoviral helper plasmid that encodes for the adenoviral E2a, E4, and VA helper genes (Grimm et al., 1999; Xiao et al., 1998a). This allows for the production of replication-deficient, wild-type-free, and adenovirus-free rAAV vectors stocks at adequate titers. To facilitate upscaling of vector production and to generate good manufacturing practice (GMP) compliant rAAV vector stocks for clinical or commercial use, several novel techniques have been proposed (Durocher et al., 2007; Zolotukhin, 2005). Such approaches are based on the generation of stably transfected producer cell lines (Blouin et al., 2004; Clark et al., 1995), suspension cell transfection and transduction techniques (Meghrous et al., 2005; Park et al., 2006; Smith et al., 2009), and even cell-free production of rAAV (Zhou and Muzyczka, 1998). Innovative purification protocols using gradient centrifugation and chromatography (Hermens et al., 1999; Zolotukhin et al., 1999) have contributed to making production and purity of stable rAAV vector stocks feasible even for large-scale production.

AAV vector transduction in most tissues is characterized by a delayed onset of gene expression. Since second strand synthesis is a rate-limiting step in AAV-mediated gene transfer, self-complementary AAV (scAAV) vectors have been generated by deletion of the D-sequence or mutation of one terminal resolution site (trs) sequence located within the ITR's. This results in the production of a high percentage of self-complementary vectors and allows for rapid and increased expression of the transgene. A serious drawback of this approach is a loss of reduced packing capacity (Duque et al., 2009; Fu et al., 2003; McCarty, 2008; McCarty et al., 2003; Zhong et al., 2004).

Vectors based on AAV-2 have been the best studied of all AAV serotype derivatives. Therefore, they have been used not only for countless experimental in vitro transductions but also for a very large number of preclinical studies in animal models. Such in vivo studies yielded promising results ranging from substantial correction to complete cure in models of hemophilia, $\alpha 1$-antitrypsin deficiency, cystic fibrosis, Duchenne muscular dystrophy, rheumatoid arthritis, and others. Furthermore, AAV has been employed for a variety of anticancer gene therapy approaches. Toward this end, common strategies are based on the delivery of cytotoxic genes, reconstitution of tumor suppressor genes, inhibition of drug resistance, immunotherapy, and antiangiogenesis. AAV vectors have also been used in clinical trials in human patients. So far, at least 70 clinical trials have been approved or completed with AAV-based vectors (Carter, 2005; Coura Rdos and Nardi, 2007; Mueller and Flotte, 2008; Park et al., 2008). However, their broad host tropism and inefficient transduction of many, if not most potential, target tissues after systemic application continue to limit the utility of AAV vector systems for the majority of potential clinical applications in which topical administration of the vector to the target tissue is not feasible.

6. AAV serotypes

AAV-2 vectors have the potential to efficiently deliver genes to a broad spectrum of dividing and nondividing cell types and tissues in vitro and in vivo, including skeletal muscle, cardiac muscle, airway epithelium, hepatocytes, brain tissue, and several cancer cell lines (Arruda et al., 2005; Bartlett et al., 1998; Fisher et al., 1997; Flotte et al., 1992; Foust et al., 2009; Hacker et al., 2005; Herzog, 2004; Kaplitt et al., 1994; Palomeque et al., 2007; Snyder et al., 1997; Vassalli et al., 2003; Xiao et al., 1996, 1997, 1998a,b).

The exploitation of the distinct tissue tropism of the various AAV serotypes provides an opportunity to improve the efficiency of gene delivery to specific target tissues. Using alternative serotypes compared to AAV-2, numerous studies have shown superior transduction rates in certain tissue. AAV-1, -3, -5, -6, -7, -8, and -9 seem to be the more efficient for transduction of various tissues including the muscle, liver, heart, the central nervous system, vascular endothelium, arthritic joints, pancreas, cochlear inner hair cells, and the retina, while AAV-5 and -6 seem to be more appropriate for transduction of the lung or the airway epithelium upon local application (Apparailly *et al.*, 2005; Blankinship *et al.*, 2004; Burger *et al.*, 2004; Chen *et al.*, 2005; Halbert and Miller, 2004; Halbert *et al.*, 2001; Limberis *et al.*, 2009; Liu *et al.*, 2005; Loiler *et al.*, 2003; Riviere *et al.*, 2006; Surace and Auricchio, 2008; Taymans *et al.*, 2007; Vandendriessche *et al.*, 2007; Wang *et al.*, 2004).

If used for systemic administration, AAV-8 and -9 are more sufficient for gene transduction than other serotypes as they can efficiently cross vascular endothelial cell barriers to transduce liver hepatocytes, cardiac and skeletal muscle cells, and various other tissues (Inagaki *et al.*, 2006; Nakai *et al.*, 2005; Pacak *et al.*, 2006; Paneda *et al.*, 2009; Wang *et al.*, 2005; Zincarelli *et al.*, 2008). AAV-9 has some particularly intriguing features if applied systemically. This serotype can deliver genes both to neuronal and nonneuronal central nervous system cells (Foust *et al.*, 2009). The latter finding is of special importance as nonneuronal cells were considered inaccessible to AAV gene transfer before. Further, work by Foust *et al.* (2009) suggested that AAV-9 has the unique property to cross the blood–brain barrier (BBB) and that the transduction of astrocytes is a receptor-mediated process that occurs via astrocytic endfeet.

Pseudotyping AAV vectors by cross-packaging of an AAV genome into the capsid of another serotype improve transduction of certain tissues *in vivo* while circumventing problems of preexisting immunity (Kwon and Schaffer, 2008; Wu *et al.*, 2006a,b). However, alternative serotypes or pseudotyped AAV vectors *per se* are not capable to mediate cell-type specific transduction of AAV or any other gene therapy vector. Some of the specificity problems may be overcome by the use of tissue-specific promoters (Halbert *et al.*, 2007; Muller *et al.*, 2006; Ruan *et al.*, 2001; Wang *et al.*, 2008). However, for many cell types, specific promoters that generate adequate expression levels are not available and do not allow for gene transduction in cells that are nonpermissive for AAV infection. Therefore, a plethora of strategies have been developed to manipulate the capsid for redirection of the vector to alternative, cell-type specific receptors.

II. CAPSID MODIFICATIONS OF AAV TO TARGET CELLULAR RECEPTORS

A. Chimeric and mosaic capsids

One possible approach to alter the tropism of AAV-derived vectors is the generation of chimeric or mosaic vectors. Chimeric vectors are generated by exchange of certain capsid domains or sequences by such domains or sequences of different serotypes (Wu et al., 2000). Mosaic capsids are generated by mixing the capsid subunits from two different serotypes or capsid mutants. This yields vectors that combine the beneficial features of the originating vector capsids that synergistically enhance gene transduction. Mosaic vectors have been generated that efficiently deliver transgenes to muscle (Hauck et al., 2003), liver (Hauck et al., 2003), or endothelial cells (Stachler and Bartlett, 2006). In addition, unexpected synergistic transduction effects on various cell lines were observed when AAV-1 subunits were mixed with such of AAV-2 or AAV-3, suggesting perspectives for vectors with novel tropism (Rabinowitz et al., 2004). Major drawbacks of mosaic or chimeric vectors are preexisting antibodies against any one of the parental serotypes and the difficulty to reproduce the exact stoichiometry of the generated capsid proteins in large-scale vector productions (Kwon and Schaffer, 2008).

B. Ligand-mediated receptor targeting

A large number of technical approaches have been chosen to generate AAV vectors displaying capsid domains for selective binding to certain target cell receptors. Basically, two different principles have been used: indirect and direct targeting. Indirect targeting uses bispecific adaptor molecules with a vector-binding and a receptor-binding moiety (Trepel et al., 2000b), whereas direct targeting implies the insertion of a target receptor-binding ligand directly into the capsid protein. While indirect targeting provides the convenience of not having to mutate the capsid genes, the direct targeting strategy is not based on conjugates and therefore has many advantages such as ease of handling, better stability in vitro and in vivo, maintenance of the small size of the vector particle and avoidance of additional immunogenicity elicited by conjugates.

The very first report on AAV targeting used an indirect approach and applied bi-specific $F(ab'\gamma)_2$ antibodies directed to AAV capsids and $\alpha_{11b}\beta3$ integrins for transduction of naturally non-AAV-permissive $\alpha_{11b}\beta3$-expressing megakaryocytic cells (Bartlett et al., 1999). Other indirect approaches used avidin-linked epidermal growth factor (EGF) or fibroblast growth factor (FGF) fusion proteins conjugated to biotinylated AAV capsids to target human ovarian cancer and megakaryocytic cell lines (Ponnazhagan et al., 2002).

In a study combining indirect targeting with capsid modification, the Z34C immunoglobulin G (IgG) binding domain of protein A was inserted into the AAV-2 capsid at position R587. Subsequently, monoclonal antibodies against CD29, CD117, or CXCR4 receptors serving as linker molecules were linked to the vector surface. Such vectors allowed for selective targeting of cell types expressing the respective receptors (Ried et al., 2002). In a related, double-targeting approach, Z34C vector particles were produced as "mosaic" particles with AAV-2 capsid proteins that allowed specific CD29 and CD117 receptors with significantly higher transduction efficiency (Gigout et al., 2005).

In the direct targeting approach, cell-specificity of the vector is mediated by a ligand coding sequence that is inserted into the VP capsid gene and presented within the viral capsid surface. Several sites in the AAV capsid have been proved to be amenable to manipulation and incorporation of targeting peptides without compromising capsid assembly and genome packing efficiency.

A set of vectors with modified tropism was generated by insertion of ligands at the N terminus of VP1 or VP2, yielding vectors targeted to CD34 expressing cells by means of a single chain antibody fragment (Yang et al., 1998), to lung epithelial cells by means of the serpin receptor ligand (Wu et al., 2000), to ovarian cancer cells via a human luteinizing hormone receptor-binding peptide (Shi et al., 2001), and to islet cells and murine hepatocytes by insertion of the ApoE ligand (Loiler et al., 2003). Although the VP1/VP2 N-termini tolerate larger insertions (up to 238 amino acids) (Lux et al., 2005; Warrington et al., 2004), they are normally located inside the native capsids (Kronenberg et al., 2005; Sonntag et al., 2006), resulting in a lag of presentation of the targeting moiety. Another disadvantage of the VP2 N-terminal epitope tagging strategy is that the HSPG binding ability of the targeted vector is not abolished, resulting in the infection of a very broad range of cell types.

Among the potential insertion sites for ligands within the AAV capsid, the most widely used one is the region surrounding amino acid positions 585/588 for several reasons. First, structural modeling revealed that a sequence inserted at this position is presented 60 times on the viral surface at the flank of each spike at the threefold axis of symmetry. Therefore, peptides inserted at this position seem to be accessible for efficient receptor–ligand interaction and do not impair capsid assembly as long as they do not exceed approximately 35 amino acids in length. Second, it has been shown that the insertion of peptides at positions adjacent to 585/588 interferes with the heparin-binding motif composed of five amino acid residues (at positions 484, 487, 532, 585, and 588) and therefore potentially abrogates the natural HSPG binding of AAV-2 capsids. Depending on the peptide sequence, this can result in a detargeting from the liver in vivo after systemic application (Michelfelder et al., 2009; Müller et al., 2003; Perabo et al., 2006). Third, AAV vectors modified at position 587 have the potential to escape the neutralizing effects of human antibodies without losing their ability to infect cells via the targeted receptors (Huttner et al., 2003). The first report on direct targeting of AAV described insertion of a 14 amino acid integrin-binding peptide at position 587, and enabling targeted gene delivery to $\alpha_V\beta5$ integrin-expressing cells (Girod et al., 1999).

The design of the peptide ligand to generate targeted vectors is not an easy task. The use of phage display libraries allowing for the identification of targeted peptide ligands even without prior knowledge of their receptors has been a significant step forward in this field (Binder et al., 2006, 2007; Tamm et al., 2003; Trepel et al., 2002, 2008). For tissue targeting in particular, major advances have been made by the exploration of organ-specific "address molecules" expressed on endothelial surfaces by *in vivo* phage display (Arap *et al.*, 2002a; Pasqualini and Ruoslahti, 1996; Trepel *et al.*, 2001, 2002, 2008; Yao *et al.*, 2005). Several peptide ligands have been identified for a variety of tissues and have subsequently been used for delivery of cytotoxic drugs or other therapeutic agents in relevant preclinical models *in vivo* (Arap *et al.*, 1998, 2002b; Ellerby *et al.*, 1999; Koivunen *et al.*, 1999; Kolonin *et al.*, 2004; Marchio *et al.*, 2004).

Such peptides were introduced successfully into the AAV capsid for redirection of vectors to cells expressing the target receptor. For example, CD13 is a key regulator of angiogenesis which is upregulated in activated blood vessels and tumor vasculature. It is the target receptor for the NGR tripeptide motif. By incorporation of an NGR-containing peptide ligand into the AAV capsid, vectors have been successfully retargeted to CD13-expressing cells (Grifman *et al.*, 2001). Other groups inserted RGD-containing peptides for targeting of $\beta1$, $\beta3$, $\alpha v\beta3$, or $\alpha v\beta5$ integrin-expressing cells (Girod *et al.*, 1999; Shi and Bartlett, 2003). The CHN sequence motif was used to target AAV to athero-sclerotic lesions *in vivo* via the membrane type-1 matrix metalloprotease (MT1-MMP) (White *et al.*, 2008). For the most peptides selected by phage display, however, the target receptor is not yet known. Nevertheless, using such "orphan" peptide ligands, AAV vectors have been successfully retargeted to endothelial cells *in vitro* (Nicklin *et al.*, 2001), to endothelial cells (White *et al.*, 2004), to lung and brain vasculature (Work *et al.*, 2006), and to cardiac tissue (Yu *et al.*, 2009) (Table 2.1).

III. SELECTING NOVEL AAV VECTORS FROM COMBINATORIAL LIBRARIES

A. Principle of random AAV-display libraries

As described above, detailed functional and structural examination of AAV serotypes has yielded gene delivery vectors with novel tropism and functional properties. However, the functional diversity of serotypes is limited. Therefore, it is pertinent to develop complementary vector engineering tools that can create novel vehicles with a desired set of properties. Even though rational approaches to designing AAV vectors that target specific cells, for example, by insertion of targeting ligands, have been explored, such approaches are often limited by inadequate knowledge of the receptors expressed on the target cell surfaces. In the absence of ligand–receptor information, targeting ligands can be identified through the use of random peptide libraries displayed on the surface of phage. As detailed in the previous paragraph, peptide ligands isolated by phage display

Table 2.1. Peptide Ligands Genetically Inserted into the AAV-2 Capsid to Enhance Gene Delivery in Barious Cells and Tissues

Selection by	Target receptors, cells, or tissues	Peptide insertion at AAV capsid positions 587/588[a]	Transduction of	References
Phage display	$\beta 1, \beta 3$ integrin	AGTFALRGDNPQG	B16F10, RN22	Girod et al. (1999)
	$\alpha v \beta 3$ and $\alpha v \beta 5$ integrin	CDCRGDCFC	HeLa, K562, Raji, SKOV-3, local in vivo	Shi and Bartlett (2003)
	CD13	NGRAHA	Sarcoma cell lines	Grifman et al. (2001)
	HUVEC cells	SIGYPLP	HUVEC, HSVEC	Nicklin et al. (2001)
	HUVEC cells	MTPFPTSNEANLGGGS	HUVEC, venous endothelial cells i.v.	White et al. (2004)
	Human saphenous vein and arterial SMC	EYHHYNK	Human saphenous vein and arterial SMC	Work et al. (2004)
	MT1-MMP	CNHRYMQMC	Atherosclerotic lesions i.v.	White et al. (2008)
	Brain endothelium i.v.	QPEHSST	Brain endothelium i.v.	Work et al. (2006)
	Lung endothelium i.v.	VNTANST	Lung endothelium i.v.	Work et al. (2006)
	Muscle i.v.	ASSLNIA	Cardiac muscle, striated muscle i.v.	Yu et al. (2009)
AAV display	Mec1	RSNAVVP	Mec1	Perabo et al. (2003)
	PBPC	NRTWEQQ	Various leukemia cell lines, PBPC	Sellner et al. (2008)
	Kasumi-1	NQVGSWS	Various hematopoietic cancer cells	Michelfelder et al. (2007)
	K562	EARVRPP	Leukemia cell lines, PBPC	Stiefelhagen et al. (2008)
	Coronary artery endothelial cells	NSSRDLG	Coronary artery endothelial cells, heart i.v.	Müller et al. (2003)
	HSaVEC	NDVRAVS	HSaVEC	Waterkamp et al. (2006)
	HSaVEC	NDVRSAN	HSaVEC	Waterkamp et al. (2006)
	Calu6	VTAGRAP	Calu6	Waterkamp et al. (2006)
	PC3	DLSNLTR	PC3	Waterkamp et al. (2006)
	M07e	RGDAVGV	M07e	Perabo et al. (2003)
	Primary PymT tumor cells	RGDLGLS	Primary PymT tumor cells, MCF-7	Michelfelder et al. (2009)
	Coronary artery endothelial cells	PRSVTVP	Coronary artery endothelial cells	Müller et al. (2003)
	PymT tumor tissue i.v.	DLGSARA	Tumor, liver, heart, lung, brain, muscle i.v.	Michelfelder et al. (2009)
	PymT tumor tissue i.v.	ESGLSQS	Tumor, heart i.v.	Michelfelder et al. (2009)
	Lung tissue i.v.	PRSTSDP	Lung, liver, heart, kidney, brain, muscle i.v.	Michelfelder et al. (2009)
AAV shuffling	Lung tissue after topic application	NSSRSLG	Lung, alveolar Type-2 cells topic; liver i.v.	Grimm et al. (2008)
AAV display	Lung tissue after topic application	MVNNFEW	Lung, alveolar macrophages; liver i.v.	Grimm et al. (2008)

[a]Single letter amino acid code.

libraries can then be incorporated into the AAV capsid, and in many cases, provided targeting to desired cell types. The success rate of this approach varies considerably, however. Our own experience has been that only a minority of selected peptide ligands function equally well in targeted phage particles and in modified AAV capsids. One of the several problems accounting for this observation may be that the phage-derived peptides were selected only for cell or receptor binding but not for the subsequent posttargeting cell entry which is required for gene transfer. In addition, the structural context the peptides are displayed is probably crucial. The binding properties of a peptide ligand may change unpredictably when it is incorporated into a virus capsid protein. For instance, the peptide might be subjected to structural constraints not present in the phage capsid which was initially used for selection of the ligand from the random library. Taking such limitations into account, we and others have developed random peptide-display libraries based on the gene therapy vector capsid itself for AAVs (Müller et al., 2003; Perabo et al., 2003) and later, for retroviruses (Bupp and Roth, 2003; Hartl et al., 2005; Khare et al., 2003a,b). Peptide ligands binding specifically to a cell type within the context of the specific viral capsid protein can therefore be selected. In the context of AAV, such libraries were termed random AAV display peptide libraries. Incubation of such AAV libraries on target cells leads to uptake of those viral particles that display a suitable peptide ligand binding to a target cell receptor. The internalized AAV particles can be amplified in the target cells by superinfection with adenoviral helper viruses and reapplied to the target cells for a refined screening procedure ("round of selection"). This principle is illustrated in Fig. 2.2. Thus, selection of AAV-display libraries yields only viral vectors that are assembled and produced—display the unique targeting moiety in the structural context of the AAV capsid—and are also able to enter and successfully transduce the target cell. Using this technology, vectors were isolated that specifically and efficiently transduced the cell types they were selected on, which is described in detail below (Table 2.1).

B. Different AAV peptide library technologies and further improvements

Initially, two different AAV library systems were introduced. Both systems were successfully used to select AAV particles able to transduce cell types which are otherwise barely susceptible to AAV-2 gene delivery (Müller et al., 2003; Perabo et al., 2003). Both strategies are based on the production of a plasmid library as a first step but they differ significantly in the production strategy of the virus particle library. Perabo et al. (2003) used direct transfection of the plasmid library into the cells producing the final AAV display peptide library, which is a time saving method but also has important disadvantages. In contrast,

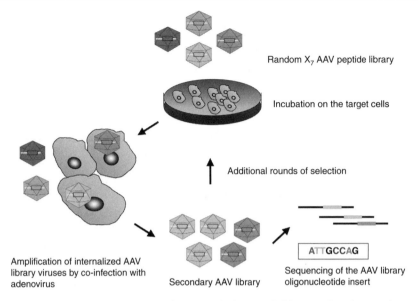

Random X₇ AAV peptide library

Incubation on the target cells

Additional rounds of selection

Amplification of internalized AAV
library viruses by co-infection with
adenovirus   Secondary AAV library

Sequencing of the AAV library
oligonucleotide insert

ATTGCCAG

Figure 2.2. Principle of selection of random AAV display peptide libraries. The selection of AAV
peptide libraries is based on the amplification of viruses that are taken up by target cells
by means of the peptide displayed on their surface. A random AAV peptide library is
incubated on the target cells of interest. Bound and internalized library viruses are
subsequently amplified by coinfection with adenovirus and used for additional rounds
of selection. Enriched peptide insertions are analyzed by DNA sequencing of recovered
clones. (See Page 2 in Color Section at the back of the book.)

Müller *et al.* (2003) used the AAV plasmid libraries to produce so-called "AAV
library transfer shuttles" consisting of mosaic capsids with wild-type and library-
derived capsid proteins, which were then used in a subsequent step to generate
the final random AAV display peptide library (Fig. 2.3). The final vector library
was produced by infection of producer cells at a very low multiplicity of infection
(usually approximately one replicative unit per producer cell; Fig. 2.3). This
minimizes production of capsids with multiple different peptides and maximizes
the likelihood that the packaged *cap* gene truly encodes for the peptide that is
displayed on the virus particle. It is expected that the functional diversity and
utility of such libraries, even though they are much more labor-intensive to
produce, is superior to the single-step system. Accordingly, a much greater
diversity of clones with multiple variants of a clear-cut peptide motif were selected
and enriched from round to round with such libraries (Müller *et al.*, 2003), while in
the "single-step library" selections, one or two dominant clones were selected and a
peptide sequence motif could not be seen (Perabo *et al.*, 2003). Nevertheless, both

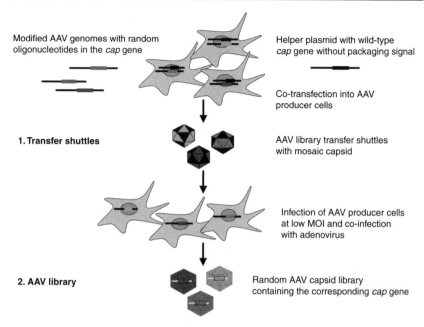

Modified AAV genomes with random oligonucleotides in the *cap* gene

Helper plasmid with wild-type *cap* gene without packaging signal

Co-transfection into AAV producer cells

1. Transfer shuttles

AAV library transfer shuttles with mosaic capsid

Infection of AAV producer cells at low MOI and co-infection with adenovirus

2. AAV library

Random AAV capsid library containing the corresponding *cap* gene

Figure 2.3. Three-step system for the generation of random AAV display peptide libraries. In the first step, a plasmid library is cloned. In the second step, AAV library transfer shuttles with mosaic capsids, harboring library *cap* genes, are generated by cotransfection of mutated *cap* genes flanked by ITRs and wild-type *cap* genes lacking the ITRs. AAV producer cells are infected with AAV transfer shuttles at a low MOI to achieve a presumed uptake and propagation of only one library genome per cell to ensure that the mutant capsid genome encodes for the displayed capsid protein. (See Page 2 in Color Section at the back of the book.)

systems work well *in vitro* and the potential functional superiority of one system over another—at the price of considerably more effort—may become more apparent in *in vivo* selections.

One drawback of the three-step procedure is that during the AAV library shuttle production (second step of the production process), both wild-type and library AAV genomes are present in the producer cell. While wild-type genomes only serve as a helper to produce capsids that are partially made up by wild-type protein, they should not be packaged in the shuttles as they lack the ITR packaging signals. Yet, it has been clearly shown that packaged wild-type genomes are indeed detectable in the shuttles, which is most likely due to homologous recombination between wild-type and library genomes within the shuttle-producing cells. As a result of wild-type genomes being packaged in the shuttles, the final AAV library will also partially consist of wild-type AAV particles. Such wild-type contamination does not play a major role in selections on cells that are not susceptible to wild-type AAV transduction. However, it is counterproductive in selections on wild-type-susceptible cells as wild-type AAV always competes with potentially suitable AAV library clones and selections can be more difficult or impossible (Michelfelder and Trepel, unpublished observation). Therefore, Waterkamp et al. (2006) developed an AAV library production system which is based on the three-step method of Müller et al. (2003). This system uses synthetic wild-type like genomes with alternative codon usage so as to completely avoid the possibility of homologous recombination and still maintain the complexity of the library (Waterkamp et al., 2006). Consequently, AAV libraries completely free of wild-type AAV were produced and yielded superior selection results compared to the first-generation libraries (Waterkamp et al., 2006).

C. AAV peptide library screenings *in vitro*

The introductory report on AAV libraries using the three-step production technology described screenings of AAV libraries on human coronary artery endothelial cells (Müller et al., 2003). Striking peptide sequence motifs were selected which were reproducible upon selection with two separately produced libraries. The selected peptides in AAV capsids enhanced AAV-mediated transduction in coronary endothelial cells but not in control nonendothelial cells (Müller et al., 2003).

Using the single-step AAV library production method, Perabo et al. (2003) described AAV library selections on myeloid and lymphoid leukemia cells that were resistant to infection by wild-type AAV. After six rounds of selection on different cell lines, infectious mutants were harvested which transduced target cells with an up to 100-fold increased efficiency in a receptor-specific manner and independently of the primary receptor for wild-type AAV, HSPG (Perabo et al., 2003). Michelfelder et al. (2007) also reported on AAV selections on myeloid leukemia cells. Like in all other work published so far which used the three-step AAV library production procedure, the peptide inserts of the enriched capsid mutants shared a common sequence motif which was selected independently on several different leukemia cell lines. Recombinant targeted vectors displaying the selected peptides transduced the target leukemia cells up to 500-fold more efficiently compared to AAV vectors with control peptide inserts. The described AAV targeting peptide with the sequence NQVGSWS was able to overcome resistance to AAV transduction in a variety of hematopoietic cancer cell types, whereas normal (i.e., noncancerous) peripheral blood mononuclear cells and CD34+ hematopoietic progenitor cells were not transduced. Consequently, NQVGSWS AAV vectors harboring a suicide gene conferred selective killing to leukemia cells, but not to control cells

(Michelfelder *et al.*, 2007). In view of the difficulty of gene delivery to hemato-poietic cancer cells (Kohlschutter *et al.*, 2008; Trepel *et al.*, 1997), this finding is of particular importance.

Using the improved wild-type AAV-free library system described above, Waterkamp *et al.* (2006) selected on five different cell types including lung cancer cells, prostate cancer cells, primary venous endothelial cells, dendritic cells, and cardiomyoblasts. Highly specific peptide sequence motifs were selected that provided a considerable increase in gene transduction over vectors carrying wild-type AAV capsids (Waterkamp *et al.*, 2006).

It is quite remarkable, that the selection of AAV display peptide libraries on all of the cell types mentioned above yielded unique, cell-type specific peptide motifs that could be reproducibly isolated from independent library productions. As it is highly likely that there are multiple kinds of receptors on the target cells that potentially mediate binding of AAV library particles, such restriction to one particular peptide motif is striking. It may reinforce suggest that the selection process not only enriches for capsids that efficiently bind the target cell, but rather for peptides that are also compatible with suitable downstream processing leading to virus internalization, endosomal escape, nuclear transport, uncoating, and transgene expression. It appears possible that there is only a limited number of receptors per target cell that meet these requirements. Thus, library selections may end up with capsids mainly targeted to such a receptor and not any other receptor which may only be compatible with binding, explaining for the homogeneity of the recovered peptide sequences.

D. AAV peptide library screenings *in vivo*

Targeting viral vectors to certain tissues *in vivo* has been a major challenge in gene therapy. As described above, cell type-directed vector capsids can be selected from random peptide libraries displayed on viral capsids *in vitro* but for a long time, this system could not easily be translated to *in vivo* applications. Based on a capsid mutant selected by DNA family shuffling, Grimm *et al.* (2008) generated a random AAV display peptide library and used it for *in vivo* selection of AAV capsids transducing the lung. The selections were done by topical applications of the library along with a helper adenovirus through the airways. After the second round of selection, one single AAV peptide mutant was enriched. Consequently, a sequence motif was not observed. AAV vectors carrying capsids with this selected peptide indeed mediated gene expression in distinct alveolar cells after topic application (Grimm *et al.*, 2008). In contrast to this study, Michelfelder *et al.* (2009) selected random AAV display peptide libraries *in vivo* after *systemic* administration. A novel, PCR-based amplification protocol for AAV-displayed peptide libraries was developed. This is a salient prerequisite for systemic *in vivo* selections, because (i) helper viruses such as adenovirus otherwise used for AAV library amplification are poorly tolerated in higher doses after intravenous injection, and (ii) helper viruses will not necessarily infect the cells and tissues targeted by the AAV library particles. Building on this PCR-based library amplification technique, selections of AAV peptide libraries *in vivo* in living mice after intravenous administration were performed using tumor and lung tissue as prototype targets. After several rounds of selection, distinct sequence motifs for both tissues were enriched. The selected clones indeed conferred gene expression in the target tissue while gene expression was undetectable in animals injected with control vectors. Importantly, however, all of the vectors selected for tumor transduction also transduced heart tissue and the vectors selected for lung transduction also transduced a number of other tissues, particularly and invariably the heart. This may imply that diminishing binding to the so-called primary AAV receptor HSPG by peptide insertion in capsid position 588 is necessary but not sufficient to achieve truly tissue-specific transgene expression (Michelfelder *et al.*, 2009). While the approach presented in this work does not yield vectors whose expression is confined to only one target tissue, it provides a very useful tool for *in vivo* tissue transduction when expansion rather than restriction of AAV tropism to the tissue of interest is needed.

E. Other approaches for selection of tropism-modified AAV: Error-prone PCR, DNA shuffling, and multispecies libraries as alternative strategies

Directed evolution has been proved as a useful tool to generate enzymes with new catalytic properties (May *et al.*, 2000; Stemmer, 1994), antibodies with optimized binding affinity (Boder and Wittrup, 1997), or retroviruses with novel properties (Soong *et al.*, 2000). Maheshri *et al.* (2006) have built upon this work to develop a high-throughput method for the generation of rAAV vectors with altered receptor-binding properties, improved transduction profiles, and the ability to evade neutralizing antibodies both *in vitro* and *in vivo*. A directed evolution approach was applied involving the generation of large mutant capsid libraries with point mutations randomly distributed throughout the AAV-2 capsid proteins VP1-3 by error-prone PCR. The library was selected on AAV-2-susceptible cells in the presence of neutralizing antiserum and clones evading neutralizing antibodies were recovered which delivered genes more efficiently than wild-type AAV-2 in the presence of anti-AAV serum *in vitro* and *in vivo* (Maheshri *et al.*, 2006).

    An alternative strategy of directed evolution was used by Grimm *et al.* (2008). In an attempt to merge desirable qualities of multiple AAV serotypes, this group used an adapted DNA family shuffling method to create a library of hybrid capsids from eight different serotypes. Selection of this library on human hepatocytes in the presence of human antisera enriched a hybrid from a single serotype 2/8/9 chimera, which differed from its closest natural relative AAV-2 by 60 capsid amino acids. Compared to the eight confounding natural AAV serotypes, this selected AAV capsid chimera conferred superior transduction capacity *in vitro* and it greatly surpassed AAV-2 in livers *in vivo* in a mouse model (Grimm *et al.*, 2008).

To engineer gene vectors that target striated muscles after systemic delivery *in vivo*, Yang *et al.* (2009) combined such serotype-shuffling approaches with the *in vivo* selection method of peptide libraries after systemic injection by Michelfelder *et al.* (2009). A random library of AAV was constructed by shuffling the capsid genes of AAV serotypes 1–9 and it was screened for muscle-targeting capsids *in vivo* in mice. A capsid hybrid of AAV-1, -6, -7, and -8 was retrieved, based on its high frequency in the muscle and low frequency in the liver. This clone showed a surprising *in vivo* transduction pattern. While gene transfer in nonmuscle tissues such as the liver was dramatically reduced, the clone transduced both heart and muscle very efficiently. In fact, the myocardium showed the highest gene expression among all tissues tested in two different rodent models after systemic administration, even though the heart was not the primary target of selection (Yang *et al.*, 2009). This finding is consistent with the results by Michelfelder *et al.* (2009), who observed predominant heart transduction after systemic injection of AAV library mutants which were primarily selected for transduction of tumor or lung tissue.

F. Ligand-directed AAV-phage hybrid vectors for targeted transgene delivery

Selecting peptide ligands for gene vector targeting in the structural context of the vector capsid has unique advantages as detailed above. A drawback of this methodology is, however, that it is very labor-intensive, especially compared to the use of random phage display libraries. For instance, cloning and production of a phage library can be accomplished within approximately 2–3 working weeks while it usually takes 4–6 weeks to clone and produce an AAV library. Also, one selection round with a phage library can usually be done in 1–2 days while a PCR-based selection round with an AAV library takes approximately 2–3 weeks. Furthermore, while there are several AAV vectors selected from AAV libraries which can transduce the cell type of interest, most of them have not been characterized as to which cellular receptor they bind to and are internalized by. Most peptide-displaying phage clones, in contrast, have been selected for binding to a specific receptor, or, if they were selected *in vivo*, their targeted receptor was subsequently identified (Arap *et al.*, 2002a; Jäger *et al.*, 2007; Kolonin *et al.*, 2004; Pasqualini *et al.*, 2000; Rajotte and Ruoslahti, 1999). In contrast to conventional eukaryotic vectors like the ones based on AAV, however, bacteriophage species (as "prokaryotic vectors") have mostly been considered unsuitable vehicles for transduction of mammalian cells. On the other hand, a potential advantage of phage-based vectors is that they have no tropism for mammalian cells (Barrow and Soothill, 1997) which must be neutralized for retargeting. Although phage vectors have even been used for transduction of eukaryotic cells (Larocca *et al.*, 1999; Piersanti *et al.*, 2004; Poul and Marks, 1999), inefficient transduction and immunogenicity have remained major obstacles for phage vectors preventing broader application. Therefore, it has been a major advance that both approaches, the use of phage-derived peptides for targeted gene delivery *and* the use of peptides for vector targeting in the structural protein context they have been selected in, can now be combined. Hajitou *et al.* (2006, 2007) have generated a novel hybrid vector containing genetic *cis*-elements from AAV-2 and of a single-stranded M13 bacteriophage derivative. This hyprid was termed AAV phage (AAVP) (Hajitou *et al.*, 2006). Incorporation of AAV-inverted terminal repeats into the phage transgene cassette resulted in improved intracellular fate of the delivered transgene (Hajitou *et al.*, 2006). A targeted AAVP prototype has been established, based on a phage that targets αv integrins in tumor blood vessels (Pasqualini *et al.*, 1997). AAVP-mediated tissue-specific ligand-directed transduction *in vivo* after systemic administration of the vector. This system was successfully used for *in vivo* imaging and suicide gene therapy in several tumor models (Hajitou *et al.*, 2006, 2008; Soghomonyan *et al.*, 2007; Trepel *et al.*, 2009).

IV. PERSPECTIVES

The multitude of approaches for AAV targeting is based on the assumption that differences in capsid sequence (and hence structure) can affect binding to cell surface receptors, intracellular trafficking, nuclear transport, uncoating, and even evasion of the AAV-immune responses, thus circumventing neutralizing immunity in the human population and allowing readministration.

   Combinatorial and serotype-shuffled capsid library techniques have proved to be powerful strategies in establishing retargeted AAV vectors. Random peptide libraries expressed on AAV instead of using peptide ligands derived from other libraries for AAV targeting may be a substantial improvement since the targeting peptide is displayed in the appropriate structural context of the AAV capsid. In the future, it will be of paramount importance to determine how capsid modifications affect the uptake and intracellular trafficking of AAV vectors, as improper routing or processing of vectors may significantly affect their success. Even though there is convincing evidence that insertion of targeting ligands at AAV-2 capsid position R588 abrogates the natural tropism to the primary attachment receptor HSPG, it is unclear whether the natural intracellular trafficking pathways of AAV-2 are maintained with modified vectors. Also, it will be important to determine the role of additional attachment and/or homing receptors for AAV which may mediate *in vivo* tissue tropism since *in vivo* selections of AAV libraries yielded capsids that target the tissue of interest but invariably also target heart tissue. A combination of libraries harboring mutations and insertions at multiple sites relevant to intracellular processing may further improve transduction efficacy and specificity.

As the number of targeting strategies increases, there will be many interesting ways to combine the utility of multiple approaches. The future designer AAV vectors may likely be products of the synergistic combinations of tools extracted from natural and directed evolution, as well as rational modification. The constant development of new and versatile methods to alter the tropism of AAV may ultimately make clinical applicability of systemic gene therapy a reality.

References

Akache, B., Grimm, D., *et al.* (2006a). The 37/67-kilodalton laminin receptor is a receptor for adeno-associated virus serotypes 8, 2, 3, and 9. *J. Virol.* **80**, 9831–9836.

Akache, B., Grimm, D., Shen, X., Fuess, S., Yant, S., Glazer, D., Park, J., and Kay, M. (2006b). A two-hybrid screen identifies cathepsins B and L as uncoating factors for adeno-associated virus 2 and 8. *Mol. Ther.* 330–339.

Apparailly, F., *et al.* (2005). Adeno-associated virus pseudotype 5 vector improves gene transfer in arthritic joints. *Hum. Gene Ther.* **16**, 426–434.

Arap, W., Pasqualini, R., and Ruoslahti, E. (1998). Cancer treatment by targeted drug delivery to tumor vasculature in a mouse model. *Science (New York, NY)* **279**, 377–380.

Arap, W., *et al.* (2002a). Steps toward mapping the human vasculature by phage display. *Nat. Med.* **8**, 121–127.

Arap, W., *et al.* (2002b). Targeting the prostate for destruction through a vascular address. *Proc. Natl. Acad. Sci. USA* **99**, 1527–1531.

Arruda, V. R., *et al.* (2005). Regional intravascular delivery of AAV-2-F.IX to skeletal muscle achieves long-term correction of hemophilia B in a large animal model. *Blood* **105**, 3458–3464.

Asokan, A., Hamra, J. B., Govindasamy, L., Agbandje-McKenna, M., and Samulski, R. J. (2006). Adeno-associated virus type 2 contains an integrin alpha5beta1 binding domain essential for viral cell entry. *J. Virol.* **80**, 8961–8969.

Bantel-Schaal, U., Braspenning-Wesch, I., and Kartenbeck, J. (2009). Adeno-associated virus type 5 exploits two different entry pathways in human embryo fibroblasts. *J. Gen. Virol.* **90**, 317–322.

Barrow, P. A., and Soothill, J. S. (1997). Bacteriophage therapy and prophylaxis: Rediscovery and renewed assessment of potential. *Trends Microbiol.* **5**, 268–271.

Bartlett, J. S., Samulski, R. J., and McCown, T. J. (1998). Selective and rapid uptake of adeno-associated virus type 2 in brain. *Hum. Gene Ther.* **9**, 1181–1186.

Bartlett, J. S., Kleinschmidt, J., Boucher, R. C., and Samulski, R. J. (1999). Targeted adeno-associated virus vector transduction of nonpermissive cells mediated by a bispecific F(ab' gamma)(2) antibody. *Nat. Biotechnol.* **17**, 181–186.

Bartlett, J. S., Wilcher, R., and Samulski, R. J. (2000). Infectious entry pathway of adeno-associated virus and adeno-associated virus vectors. *J. Virol.* **74**, 2777–2785.

Binder, M., Otto, F., Mertelsmann, R., Veelken, H., and Trepel, M. (2006). The epitope recognized by rituximab. *Blood* **108,** 1975–1978.

Binder, M., *et al.* (2007). Identification of their epitope reveals the structural basis for the mechanism of action of the immunosuppressive antibodies basiliximab and daclizumab. *Cancer Res.* **67,** 3518–3523.

Blankinship, M. J., *et al.* (2004). Efficient transduction of skeletal muscle using vectors based on adeno-associated virus serotype 6. *Mol. Ther.* **10,** 671–678.

Bleker, S., Sonntag, F., and Kleinschmidt, J. A. (2005). Mutational analysis of narrow pores at the fivefold symmetry axes of adeno-associated virus type 2 capsids reveals a dual role in genome packaging and activation of phospholipase A2 activity. *J. Virol.* **79,** 2528–2540.

Bleker, S., Pawlita, M., and Kleinschmidt, J. A. (2006). Impact of capsid conformation and Rep-capsid interactions on adeno-associated virus type 2 genome packaging. *J. Virol.* **80,** 810–820.

Blouin, V., *et al.* (2004). Improving rAAV production and purification: Towards the definition of a scaleable process. *J. Gene Med.* **6**(Suppl. 1), S223–228.

Boder, E. T., and Wittrup, K. D. (1997). Yeast surface display for screening combinatorial polypeptide libraries. *Nat. Biotechnol.* **15,** 553–557.

Boyle, M. P., *et al.* (2006). Membrane-associated heparan sulfate is not required for rAAV-2 infection of human respiratory epithelia. *Virol. J.* **3**(29).

Bupp, K., and Roth, M. J. (2003). Targeting a retroviral vector in the absence of a known cell-targeting ligand. *Hum. Gene Ther.* **14,** 1557–1564.

Burger, C., *et al.* (2004). Recombinant AAV viral vectors pseudotyped with viral capsids from serotypes 1, 2, and 5 display differential efficiency and cell tropism after delivery to different regions of the central nervous system. *Mol. Ther.* **10,** 302–317.

Carter, B. J. (2005). Adeno-associated virus vectors in clinical trials. *Hum. Gene Ther.* **16,** 541–550.

Chen, S., *et al.* (2005). Efficient transduction of vascular endothelial cells with recombinant adeno-associated virus serotype 1 and 5 vectors. *Hum. Gene Ther.* **16,** 235–247.

Clark, K. R., Voulgaropoulou, F., Fraley, D. M., and Johnson, P. R. (1995). Cell lines for the production of recombinant adeno-associated virus. *Hum. Gene Ther.* **6,** 1329–1341.

Coura Rdos, S., and Nardi, N. B. (2007). The state of the art of adeno-associated virus-based vectors in gene therapy. *Virol. J.* **4,** 99.

Ding, W., Zhang, L., Yan, Z., and Engelhardt, J. F. (2005). Intracellular trafficking of adeno-associated viral vectors. *Gene Ther.* **12,** 873–880.

Di Pasquale, G., *et al.* (2003). Identification of PDGFR as a receptor for AAV-5 transduction. *Nat. Med.* **9,** 1306–1312.

Duan, D., Yue, Y., Yan, Z., Yang, J., and Engelhardt, J. F. (2000). Endosomal processing limits gene transfer to polarized airway epithelia by adeno-associated virus. *J. Clin. Invest.* **105,** 1573–1587.

Duque, S., *et al.* (2009). Intravenous administration of self-complementary AAV9 enables transgene delivery to adult motor neurons. *Mol. Ther.* **17,** 1187–1196.

Durocher, Y., *et al.* (2007). Scalable serum-free production of recombinant adeno-associated virus type 2 by transfection of 293 suspension cells. *J. Virol. Methods* **144,** 32–40.

Ellerby, H. M., *et al.* (1999). Anti-cancer activity of targeted pro-apoptotic peptides. *Nat. Med.* **5,** 1032–1038.

Fisher, K. J., *et al.* (1997). Recombinant adeno-associated virus for muscle directed gene therapy. *Nat. Med.* **3,** 306–312.

Flotte, T. R., *et al.* (1992). Gene expression from adeno-associated virus vectors in airway epithelial cells. *Am. J. Respir. Cell Mol. Biol.* **7,** 349–356.

Foust, K. D., *et al.* (2009). Intravascular AAV9 preferentially targets neonatal neurons and adult astrocytes. *Nat. Biotechnol.* **27,** 59–65.

Fu, H., *et al.* (2003). Self-complementary adeno-associated virus serotype 2 vector: Global distribution and broad dispersion of AAV-mediated transgene expression in mouse brain. *Mol. Ther.* **8**, 911–917.

Gao, G. P., *et al.* (2002). Novel adeno-associated viruses from rhesus monkeys as vectors for human gene therapy. *Proc. Natl. Acad. Sci. USA* **99**, 11854–11859.

Gigout, L., *et al.* (2005). Altering AAV tropism with mosaic viral capsids. *Mol. Ther.* **11**, 856–865.

Girod, A., *et al.* (1999). Genetic capsid modifications allow efficient re-targeting of adeno-associated virus type 2. *Nat. Med.* **5**, 1052–1056.

Girod, A., *et al.* (2002). The VP1 capsid protein of adeno-associated virus type 2 is carrying a phospholipase A2 domain required for virus infectivity. *J. Gen. Virol.* **83**, 973–978.

Goncalves, M. A. (2005). Adeno-associated virus: From defective virus to effective vector. *Virol. J.* 2(43).

Govindasamy, L., *et al.* (2006). Structurally mapping the diverse phenotype of adeno-associated virus serotype 4. *J. Virol.* **80**, 11556–11570.

Grieger, J. C., and Samulski, R. J. (2005). Packaging capacity of adeno-associated virus serotypes: Impact of larger genomes on infectivity and postentry steps. *J. Virol.* **79**, 9933–9944.

Grieger, J. C., Snowdy, S., and Samulski, R. J. (2006). Separate basic region motifs within the adeno-associated virus capsid proteins are essential for infectivity and assembly. *J. Virol.* **80**, 5199–5210.

Grieger, J. C., Johnson, J. S., Gurda-Whitaker, B., Agbandje-McKenna, M., and Samulski, R. J. (2007). Surface-exposed adeno-associated virus Vp1-NLS capsid fusion protein rescues infectivity of noninfectious wild-type Vp2/Vp3 and Vp3-only capsids but not that of fivefold pore mutant virions. *J. Virol.* **81**, 7833–7843.

Grifman, M., *et al.* (2001). Incorporation of tumor-targeting peptides into recombinant adeno-associated virus capsids. *Mol. Ther.* **3**, 964–975.

Grimm, D., and Kay, M. A. (2003). From virus evolution to vector revolution: Use of naturally occurring serotypes of adeno-associated virus (AAV) as novel vectors for human gene therapy. *Curr. Gene Ther.* **3**, 281–304.

Grimm, D., *et al.* (1999). Titration of AAV-2 particles via a novel capsid ELISA: Packaging of genomes can limit production of recombinant AAV-2. *Gene Ther.* **6**, 1322–1330.

Grimm, D., *et al.* (2008). *In vitro* and *in vivo* gene therapy vector evolution via multispecies interbreeding and re-targeting of adeno-associated viruses. *J. Virol.* **82**, 5887–5911.

Hacker, U. T., *et al.* (2005). Adeno-associated virus serotypes 1 to 5 mediated tumor cell directed gene transfer and improvement of transduction efficiency. *J. Gene Med.* **7**, 1429–1438.

Hajitou, A., *et al.* (2006). A hybrid vector for ligand-directed tumor targeting and molecular imaging. *Cell* **125**, 385–398.

Hajitou, A., *et al.* (2007). Design and construction of targeted AAVP vectors for mammalian cell transduction. *Nat. Protoc.* **2**, 523–531.

Hajitou, A., *et al.* (2008). A preclinical model for predicting drug response in soft-tissue sarcoma with targeted AAVP molecular imaging. *Proc. Natl. Acad. Sci. USA* **105**, 4471–4476.

Halbert, C. L., and Miller, A. D. (2004). AAV-mediated gene transfer to mouse lungs. *Methods Mol. Biol. (Clifton, NJ)* **246**, 201–212.

Halbert, C. L., Allen, J. M., and Miller, A. D. (2001). Adeno-associated virus type 6 (AAV6) vectors mediate efficient transduction of airway epithelial cells in mouse lungs compared to that of AAV2 vectors. *J. Virol.* **75**, 6615–6624.

Halbert, C. L., Lam, S. L., and Miller, A. D. (2007). High-efficiency promoter-dependent transduction by adeno-associated virus type 6 vectors in mouse lung. *Hum. Gene Ther.* **18**, 344–354.

Handa, A., Muramatsu, S., Qiu, J., Mizukami, H., and Brown, K. E. (2000). Adeno-associated virus (AAV)-3-based vectors transduce haematopoietic cells not susceptible to transduction with AAV-2-based vectors. *J. Gen. Virol.* **81**, 2077–2084.

Hansen, J., Qing, K., Kwon, H. J., Mah, C., and Srivastava, A. (2000). Impaired intracellular trafficking of adeno-associated virus type 2 vectors limits efficient transduction of murine fibroblasts. J. Virol. 74, 992–996.

Hansen, J., Qing, K., and Srivastava, A. (2001). Adeno-associated virus type 2-mediated gene transfer: Altered endocytic processing enhances transduction efficiency in murine fibroblasts. J. Virol. 75, 4080–4090.

Hartl, I., et al. (2005). Library-based selection of retroviruses selectively spreading through matrix metalloprotease-positive cells. Gene Ther. 12, 918–926.

Hauck, B., Chen, L., and Xiao, W. (2003). Generation and characterization of chimeric recombinant AAV vectors. Mol. Ther. 7, 419–425.

Hermens, W. T., et al. (1999). Purification of recombinant adeno-associated virus by iodixanol gradient ultracentrifugation allows rapid and reproducible preparation of vector stocks for gene transfer in the nervous system. Hum. Gene Ther. 10, 1885–1891.

Herzog, R. W. (2004). AAV-mediated gene transfer to skeletal muscle. Methods Mol. Biol. (Clifton, NJ) 246, 179–194.

Hunter, L. A., and Samulski, R. J. (1992). Colocalization of adeno-associated virus Rep and capsid proteins in the nuclei of infected cells. J. Virol. 66, 317–324.

Huttner, N. A., et al. (2003). Genetic modifications of the adeno-associated virus type 2 capsid reduce the affinity and the neutralizing effects of human serum antibodies. Gene Ther. 10, 2139–2147.

Inagaki, K., et al. (2006). Robust systemic transduction with AAV9 vectors in mice: Efficient global cardiac gene transfer superior to that of AAV8. Mol. Ther. 14, 45–53.

Jäger, S., et al. (2007). Leukemia-targeting ligands isolated from phage-display peptide libraries. Leukemia 21, 411–420.

Jennings, K., et al. (2005). Proteasome inhibition enhances AAV-mediated transgene expression in human synoviocytes in vitro and in vivo. Mol. Ther. 11, 600–607.

Johnson, J. S., and Samulski, R. J. (2009). Enhancement of adeno-associated virus infection by mobilizing capsids into and out of the nucleolus. J. Virol. 83, 2632–2644.

Kaludov, N., Brown, K. E., Walters, R. W., Zabner, J., and Chiorini, J. A. (2001). Adeno-associated virus serotype 4 (AAV4) and AAV5 both require sialic acid binding for hemagglutination and efficient transduction but differ in sialic acid linkage specificity. J. Virol. 75, 6884–6893.

Kaplitt, M. G., et al. (1994). Long-term gene expression and phenotypic correction using adeno-associated virus vectors in the mammalian brain. Nat. Genet. 8, 148–154.

Kashiwakura, Y., et al. (2005). Hepatocyte growth factor receptor is a coreceptor for adeno-associated virus type 2 infection. J. Virol. 79, 609–614.

Kern, A., et al. (2003). Identification of a heparin-binding motif on adeno-associated virus type 2 capsids. J. Virol. 77, 11072–11081.

Khare, P. D., Rosales, A. G., Bailey, K. R., Russell, S. J., and Federspiel, M. J. (2003a). Epitope selection from an uncensored peptide library displayed on avian leukosis virus. Virology 315, 313–321.

Khare, P. D., Russell, S. J., and Federspiel, M. J. (2003b). Avian leukosis virus is a versatile eukaryotic platform for polypeptide display. Virology 315, 303–312.

King, J. A., Dubielzig, R., Grimm, D., and Kleinschmidt, J. A. (2001). DNA helicase-mediated packaging of adeno-associated virus type 2 genomes into preformed capsids. EMBO J. 20, 3282–3291.

Kohlschutter, J., Michelfelder, S., and Trepel, M. (2008). Drug delivery in acute myeloid leukemia. Expert. Opin. Drug Deliv. 5, 653–663.

Koivunen, E., et al. (1999). Tumor targeting with a selective gelatinase inhibitor. Nat. Biotechnol. 17, 768–774.

Kolonin, M. G., Saha, P. K., Chan, L., Pasqualini, R., and Arap, W. (2004). Reversal of obesity by targeted ablation of adipose tissue. *Nat. Med.* **10,** 625–632.

Kotin, R. M., Linden, R. M., and Berns, K. I. (1992). Characterization of a preferred site on human chromosome 19q for integration of adeno-associated virus DNA by non-homologous recombination. *EMBO J.* **11,** 5071–5078.

Kronenberg, S., Bottcher, B., von der Lieth, C. W., Bleker, S., and Kleinschmidt, J. A. (2005). A conformational change in the adeno-associated virus type 2 capsid leads to the exposure of hidden VP1 N termini. *J. Virol.* **79,** 5296–5303.

Kurzeder, C., *et al.* (2007). CD9 promotes adeno-associated virus type 2 infection of mammary carcinoma cells with low cell surface expression of heparan sulphate proteoglycans. *Int. J. Mol. Med.* **19,** 325–333.

Kwon, I., and Schaffer, D. V. (2008). Designer gene delivery vectors: Molecular engineering and evolution of adeno-associated viral vectors for enhanced gene transfer. *Pharm. Res.* **25,** 489–499.

Larocca, D., *et al.* (1999). Gene transfer to mammalian cells using genetically targeted filamentous bacteriophage. *FASEB J.* **13,** 727–734.

Limberis, M. P., Vandenberghe, L. H., Zhang, L., Pickles, R. J., and Wilson, J. M. (2009). Transduction efficiencies of novel AAV vectors in mouse airway epithelium *in vivo* and human ciliated airway epithelium *in vitro*. *Mol. Ther.* **17,** 294–301.

Liu, Y., *et al.* (2005). Specific and efficient transduction of cochlear inner hair cells with recombinant adeno-associated virus type 3 vector. *Mol. Ther.* **12,** 725–733.

Lochrie, M. A., *et al.* (2006). Mutations on the external surfaces of adeno-associated virus type 2 capsids that affect transduction and neutralization. *J. Virol.* **80,** 821–834.

Loiler, S. A., *et al.* (2003). Targeting recombinant adeno-associated virus vectors to enhance gene transfer to pancreatic islets and liver. *Gene Ther.* **10,** 1551–1558.

Lux, K., *et al.* (2005). Green fluorescent protein-tagged adeno-associated virus particles allow the study of cytosolic and nuclear trafficking. *J. Virol.* **79,** 11776–11787.

Maheshri, N., Koerber, J. T., Kaspar, B. K., and Schaffer, D. V. (2006). Directed evolution of adeno-associated virus yields enhanced gene delivery vectors. *Nat. Biotechnol.* **24,** 198–204.

Marchio, S., *et al.* (2004). Aminopeptidase A is a functional target in angiogenic blood vessels. *Cancer Cell* **5,** 151–162.

May, O., Nguyen, P. T., and Arnold, F. H. (2000). Inverting enantioselectivity by directed evolution of hydantoinase for improved production of L-methionine. *Nat. Biotechnol.* **18,** 317–320.

McCarty, D. M. (2008). Self-complementary AAV vectors; advances and applications. *Mol. Ther.* **16,** 1648–1656.

McCarty, D. M., *et al.* (2003). Adeno-associated virus terminal repeat (TR) mutant generates self-complementary vectors to overcome the rate-limiting step to transduction *in vivo*. *Gene Ther.* **10,** 2112–2118.

Meghrous, J., *et al.* (2005). Production of recombinant adeno-associated viral vectors using a baculovirus/insect cell suspension culture system: From shake flasks to a 20-L bioreactor. *Biotechnol. Progr.* **21,** 154–160.

Michelfelder, S., *et al.* (2007). Vectors selected from adeno-associated viral display peptide libraries for leukemia cell-targeted cytotoxic gene therapy. *Exp. Hematol.* **35,** 1766–1776.

Michelfelder, S., *et al.* (2009). Successful expansion but not complete restriction of tropism of adeno-associated virus by *in vivo* biopanning of random virus display Peptide libraries. *PloS One* **4,** e5122.

Miller, E. B., *et al.* (2006). Production, purification and preliminary X-ray crystallographic studies of adeno-associated virus serotype 1. *Acta Crystallogr.* **62,** 1271–1274.

Mitchell, M., *et al.* (2009). Production, purification and preliminary X-ray crystallographic studies of adeno-associated virus serotype 9. *Acta Crystallogr.* **65,** 715–718.

Mori, S., Wang, L., Takeuchi, T., and Kanda, T. (2004). Two novel adeno-associated viruses from cynomolgus monkey: Pseudotyping characterization of capsid protein. *Virology* **330,** 375–383.

Mueller, C., and Flotte, T. R. (2008). Clinical gene therapy using recombinant adeno-associated virus vectors. *Gene Ther.* **15,** 858–863.

Müller, O. J., et al. (2003). Random peptide libraries displayed on adeno-associated virus to select for targeted gene therapy vectors. *Nat. Biotechnol.* **21,** 1040–1046.

Muller, O. J., et al. (2006). Improved cardiac gene transfer by transcriptional and transductional targeting of adeno-associated viral vectors. *Cardiovasc. Res.* **70,** 70–78.

Nakai, H., et al. (2005). Unrestricted hepatocyte transduction with adeno-associated virus serotype 8 vectors in mice. *J. Virol.* **79,** 214–224.

Nam, H. J., et al. (2007). Structure of adeno-associated virus serotype 8, a gene therapy vector. *J. Virol.* **81,** 12260–12271.

Nicklin, S. A., et al. (2001). Efficient and selective AAV2-mediated gene transfer directed to human vascular endothelial cells. *Mol. Ther.* **4,** 174–181.

O'Donnell, J., Taylor, K. A., and Chapman, M. S. (2009). Adeno-associated virus-2 and its primary cellular receptor–Cryo-EM structure of a heparin complex. *Virology* **385,** 434–443.

Opie, S. R., Warrington, K. H., Agbandje-McKenna, M., Zolotukhin, S., and Muzyczka, N. (2003). Identification of amino acid residues in the capsid proteins of adeno-associated virus type 2 that contribute to heparan sulfate proteoglycan binding. *J. Virol.* **77,** 6995–7006.

Pacak, C. A., et al. (2006). Recombinant adeno-associated virus serotype 9 leads to preferential cardiac transduction in vivo. *Circ. Res.* **99,** e3–e9.

Palomeque, J., et al. (2007). Efficiency of eight different AAV serotypes in transducing rat myocardium in vivo. *Gene Ther.* **14,** 989–997.

Paneda, A., et al. (2009). Effect of adeno-associated virus serotype and genomic structure on liver transduction and biodistribution in mice of both genders. *Hum. Gene Ther.* **20,** 908–917.

Park, J. Y., Lim, B. P., Lee, K., Kim, Y. G., and Jo, E. C. (2006). Scalable production of adeno-associated virus type 2 vectors via suspension transfection. *Biotechnol. Bioeng.* **94,** 416–430.

Park, K., et al. (2008). Cancer gene therapy using adeno-associated virus vectors. *Front Biosci.* **13,** 2653–2659.

Pasqualini, R., and Ruoslahti, E. (1996). Organ targeting in vivo using phage display peptide libraries. *Nature* **380,** 364–366.

Pasqualini, R., Koivunen, E., and Ruoslahti, E. (1997). Alpha v integrins as receptors for tumor targeting by circulating ligands. *Nat. Biotechnol.* **15,** 542–546.

Pasqualini, R., et al. (2000). Aminopeptidase N is a receptor for tumor-homing peptides and a target for inhibiting angiogenesis. *Cancer Res.* **60,** 722–727.

Perabo, L., et al. (2003). In vitro selection of viral vectors with modified tropism: The adeno-associated virus display. *Mol. Ther.* **8,** 151–157.

Perabo, L., et al. (2006). Heparan sulfate proteoglycan binding properties of adeno-associated virus retargeting mutants and consequences for their in vivo tropism. *J. Virol.* **80,** 7265–7269.

Piersanti, S., et al. (2004). Mammalian cell transduction and internalization properties of lambda phages displaying the full-length adenoviral penton base or its central domain. *J. Mol. Med.* **82,** 467–476.

Ponnazhagan, S., Mahendra, G., Kumar, S., Thompson, J. A., and Castillas, M., Jr. (2002). Conjugate-based targeting of recombinant adeno-associated virus type 2 vectors by using avidin-linked ligands. *J. Virol.* **76,** 12900–12907.

Poul, M. A., and Marks, J. D. (1999). Targeted gene delivery to mammalian cells by filamentous bacteriophage. *J. Mol. Biol.* **288,** 203–211.

Qing, K., et al. (1999). Human fibroblast growth factor receptor 1 is a co-receptor for infection by adeno-associated virus 2. *Nat. Med.* **5,** 71–77.

Quesada, O., et al. (2007). Production, purification and preliminary X-ray crystallographic studies of adeno-associated virus serotype 7. *Acta Crystallogr.* **63,** 1073–1076.

Rabinowitz, J. E., *et al.* (2004). Cross-dressing the virion: The transcapsidation of adeno-associated virus serotypes functionally defines subgroups. *J. Virol.* **78**, 4421–4432.

Rajotte, D., and Ruoslahti, E. (1999). Membrane dipeptidase is the receptor for a lung-targeting peptide identified by *in vivo* phage display. *J. Biol. Chem.* **274**, 11593–11598.

Ried, M. U., Girod, A., Leike, K., Buning, H., and Hallek, M. (2002). Adeno-associated virus capsids displaying immunoglobulin-binding domains permit antibody-mediated vector retargeting to specific cell surface receptors. *J. Virol.* **76**, 4559–4566.

Riviere, C., Danos, O., and Douar, A. M. (2006). Long-term expression and repeated administration of AAV type 1, 2 and 5 vectors in skeletal muscle of immunocompetent adult mice. *Gene Ther.* **13**, 1300–1308.

Ruan, H., *et al.* (2001). A hypoxia-regulated adeno-associated virus vector for cancer-specific gene therapy. *Neoplasia (New York, NY)* **3**, 255–263.

Samulski, R. J., *et al.* (1991). Targeted integration of adeno-associated virus (AAV) into human chromosome 19. *EMBO J.* **10**, 3941–3950.

Sanlioglu, S., *et al.* (2000). Endocytosis and nuclear trafficking of adeno-associated virus type 2 are controlled by rac1 and phosphatidylinositol-3 kinase activation. *J. Virol.* **74**, 9184–9196.

Schmidt, M., *et al.* (2008). Adeno-associated virus type 12 (AAV12): A novel AAV serotype with sialic acid- and heparan sulfate proteoglycan-independent transduction activity. *J. Virol.* **82**, 1399–1406.

Schnepp, B. C., Jensen, R. L., Chen, C. L., Johnson, P. R., and Clark, K. R. (2005). Characterization of adeno-associated virus genomes isolated from human tissues. *J. Virol.* **79**, 14793–14803.

Sellner, L., Stiefelhagen, M., Kleinschmidt, J. A., Laufs, S., Wenz, F., Fruehauf, S., Zeller, W. J., and Veldwijk, M. R. (2008). Generation of efficient human blood progenitor-targeted recombinant adeno-associated viral vectors (AAV) by applying an AAV random peptide library on primary human hematopoietic progenitor cells. *Exp Hematol.* **36**(8), 957–964.

Seiler, M. P., Miller, A. D., Zabner, J., and Halbert, C. L. (2006). Adeno-associated virus types 5 and 6 use distinct receptors for cell entry. *Hum. Gene Ther.* **17**, 10–19.

Shi, W., and Bartlett, J. S. (2003). RGD inclusion in VP3 provides adeno-associated virus type 2 (AAV2)-based vectors with a heparan sulfate-independent cell entry mechanism. *Mol. Ther.* **7**, 515–525.

Shi, W. F., Arnold, G. S., and Bartlett, J. S. (2001). Insertional mutagenesis of the adeno-associated virus type 2 (AAV2) capsid gene and generation of AAV2 vectors targeted to alternative cell-surface receptors. *Hum. Gene Ther.* **12**, 1697–1711.

Smith, R. H., Levy, J. R., and Kotin, R.M (2009). A simplified baculovirus-AAV expression vector system coupled with one-step affinity purification yields high-titer rAAV stocks from insect cells. *Mol. Ther.* [e pub ahead of print].

Snyder, R. O., *et al.* (1997). Persistent and therapeutic concentrations of human factor IX in mice after hepatic gene transfer of recombinant AAV vectors. *Nat. Genet.* **16**, 270–276.

Soghomonyan, S., *et al.* (2007). Molecular PET imaging of HSV1-tk reporter gene expression using [18F]FEAU. *Nat. Protoc.* **2**, 416–423.

Somia, N., and Verma, I. M. (2000). Gene therapy: Trials and tribulations. *Nat. Rev. Genet.* **1**, 91–99.

Sonntag, F., Bleker, S., Leuchs, B., Fischer, R., and Kleinschmidt, J. A. (2006). Adeno-associated virus type 2 capsids with externalized VP1/VP2 trafficking domains are generated prior to passage through the cytoplasm and are maintained until uncoating occurs in the nucleus. *J. Virol.* **80**, 11040–11054.

Soong, N. W., *et al.* (2000). Molecular breeding of viruses. *Nat. Genet.* **25**, 436–439.

Stachler, M. D., and Bartlett, J. S. (2006). Mosaic vectors comprised of modified AAV1 capsid proteins for efficient vector purification and targeting to vascular endothelial cells. *Gene Ther.* **13**, 926–931.

Stemmer, W. P. (1994). Rapid evolution of a protein *in vitro* by DNA shuffling. *Nature* **370**, 389–391.

Stiefelhagen, M., Sellner, L., Kleinschmidt, J. A., Jauch, A., Laufs, S., Wenz, F., Zeller, W. J., Fruehauf, S., and Veldwijk, M. R. (2008). Application of a haematopoetic progenitor cell-targeted adeno-associated viral (AAV) vector established by selection of an AAV random peptide library on a leukaemia cell line. *Genet Vaccines Ther* **6,** 12.

Summerford, C., and Samulski, R. J. (1998). Membrane-associated heparan sulfate proteoglycan is a receptor for adeno-associated virus type 2 virions. *J. Virol.* **72,** 1438–1445.

Summerford, C., Bartlett, J. S., and Samulski, R. J. (1999). alpha V beta 5 integrin: A co-receptor for adeno-associated virus type 2 infection. *Nat. Med.* **5,** 78–82.

Surace, E. M., and Auricchio, A. (2008). Versatility of AAV vectors for retinal gene transfer. *Vis. Res.* **48,** 353–359.

Tamm, I., *et al.* (2003). Peptides targeting caspase inhibitors. *J. Biol. Chem.* **278,** 14401–14405.

Taymans, J. M., *et al.* (2007). Comparative analysis of adeno-associated viral vector serotypes 1, 2, 5, 7, and 8 in mouse brain. *Hum. Gene Ther.* **18,** 195–206.

Thomas, C. E., Ehrhardt, A., and Kay, M. A. (2003). Progress and problems with the use of viral vectors for gene therapy. *Nat. Rev. Genet.* **4,** 346–358.

Trepel, M., *et al.* (1997). A new look at the role of p53 in leukemia cell sensitivity to chemotherapy. *Leukemia* **11,** 1842–1849.

Trepel, M., Arap, W., and Pasqualini, R. (2000a). Exploring vascular heterogeneity for gene therapy targeting. *Gene Ther.* **7,** 2059–2060.

Trepel, M., Grifman, M., Weitzman, M. D., and Pasqualini, R. (2000b). Molecular adaptors for vascular-targeted adenoviral gene delivery. *Hum. Gene Ther.* **11,** 1971–1981.

Trepel, M., Arap, W., and Pasqualini, R. (2001). Modulation of the immune response by systemic targeting of antigens to lymph nodes. *Cancer Res.* **61,** 8110–8112.

Trepel, M., Arap, W., and Pasqualini, R. (2002). In vivo phage display and vascular heterogeneity: Implications for targeted medicine. *Curr. Opin. Chem. Biol.* **6,** 399–404.

Trepel, M., Pasqualini, R., and Arap, W. (2008). Screening phage-display peptide libraries for vascular targeted peptides. *Methods Enzymol.* **445,** 83–106, Chapter 4.

Trepel, M., *et al.* (2009). A heterotypic bystander effect for tumor cell killing after adeno-associated virus/phage-mediated, vascular-targeted suicide gene transfer. *Mol. Cancer Ther.* **8,** 2383–2391.

Vandenberghe, L. H., *et al.* (2006). Heparin binding directs activation of T cells against adeno-associated virus serotype 2 capsid. *Nat. Med.* **12,** 967–971.

Vandendriessche, T., *et al.* (2007). Efficacy and safety of adeno-associated viral vectors based on serotype 8 and 9 vs. lentiviral vectors for hemophilia B gene therapy. *J. Thromb. Haemost.* **5,** 16–24.

Vassalli, G., Bueler, H., Dudler, J., von Segesser, L. K., and Kappenberger, L. (2003). Adeno-associated virus (AAV) vectors achieve prolonged transgene expression in mouse myocardium and arteries in vivo: A comparative study with adenovirus vectors. *Int. J. Cardiol.* **90,** 229–238.

Vihinen-Ranta, M., Wang, D., Weichert, W. S., and Parrish, C. R. (2002). The VP1 N-terminal sequence of canine parvovirus affects nuclear transport of capsids and efficient cell infection. *J. Virol.* **76,** 1884–1891.

Walters, R. W., *et al.* (2004). Structure of adeno-associated virus serotype 5. *J. Virol.* **78,** 3361–3371.

Wang, A. Y., Peng, P. D., Ehrhardt, A., Storm, T. A., and Kay, M. A. (2004). Comparison of adenoviral and adeno-associated viral vectors for pancreatic gene delivery in vivo. *Hum. Gene Ther.* **15,** 405–413.

Wang, Z., *et al.* (2005). Adeno-associated virus serotype 8 efficiently delivers genes to muscle and heart. *Nat. Biotechnol.* **23,** 321–328.

Wang, Y., *et al.* (2008). Potent antitumor effect of TRAIL mediated by a novel adeno-associated viral vector targeting to telomerase activity for human hepatocellular carcinoma. *J. Gene Med.* **10,** 518–526.

Warrington, K. H., *et al.* (2004). Adeno-associated virus type 2 VP2 capsid protein is nonessential and can tolerate large peptide insertions at its N terminus. *J. Virol.* **78,** 6595–6609.

Waterkamp, D. A., Muller, O. J., Ying, Y., Trepel, M., and Kleinschmidt, J. A. (2006). Isolation of targeted AAV2 vectors from novel virus display libraries. *J. Gene Med.* **8**, 1307–1319.

White, S. J., *et al.* (2004). Targeted gene delivery to vascular tissue *in vivo* by tropism-modified adeno-associated virus vectors. *Circulation* **109**, 513–519.

White, K., *et al.* (2008). Engineering adeno-associated virus 2 vectors for targeted gene delivery to atherosclerotic lesions. *Gene Ther.* **15**, 443–451.

Wistuba, A., Kern, A., Weger, S., Grimm, D., and Kleinschmidt, J. A. (1997). Subcellular compartmentalization of adeno-associated virus type 2 assembly. *J. Virol.* **71**, 1341–1352.

Wobus, C. E., *et al.* (2000). Monoclonal antibodies against the adeno-associated virus type 2 (AAV-2) capsid: Epitope mapping and identification of capsid domains involved in AAV-2-cell interaction and neutralization of AAV-2 infection. *J. Virol.* **74**, 9281–9293.

Work, L. M., Nicklin, S. A., Brain, N. J., *et al.* (2004). Development of efficient viral vectors selective for vascular smooth muscle cells. *Mol Ther* **9**, 198–208.

Work, L. M., *et al.* (2006). Vascular bed-targeted *in vivo* gene delivery using tropism-modified adeno-associated viruses. *Mol. Ther.* **13**, 683–693.

Wu, P., *et al.* (2000). Mutational analysis of the adeno-associated virus type 2 (AAV2) capsid gene and construction of AAV2 vectors with altered tropism. *J. Virol.* **74**, 8635–8647.

Wu, Z., Asokan, A., and Samulski, R. J. (2006a). Adeno-associated virus serotypes: Vector toolkit for human gene therapy. *Mol. Ther.* **14**, 316–327.

Wu, Z., Miller, E., Agbandje-McKenna, M., and Samulski, R. J. (2006b). Alpha2, 3 and alpha2, 6 N-linked sialic acids facilitate efficient binding and transduction by adeno-associated virus types 1 and 6. *J. Virol.* **80**, 9093–9103.

Xiao, X., Li, J., and Samulski, R. J. (1996). Efficient long-term gene transfer into muscle tissue of immunocompetent mice by adeno-associated virus vector. *J. Virol.* **70**, 8098–8108.

Xiao, X., Li, J., McCown, T. J., and Samulski, R. J. (1997). Gene transfer by adeno-associated virus vectors into the central nervous system. *Exp. Neurol.* **144**, 113–124.

Xiao, X., Li, J., and Samulski, R. J. (1998a). Production of high-titer recombinant adeno-associated virus vectors in the absence of helper adenovirus. *J. Virol.* **72**, 2224–2232.

Xiao, W., *et al.* (1998b). Adeno-associated virus as a vector for liver-directed gene therapy. *J. Virol.* **72**, 10222–10226.

Xiao, W., Warrington, K. H., Jr., Hearing, P., Hughes, J., and Muzyczka, N. (2002). Adenovirus-facilitated nuclear translocation of adeno-associated virus type 2. *J. Virol.* **76**, 11505–11517.

Xie, Q., *et al.* (2002). The atomic structure of adeno-associated virus (AAV-2), a vector for human gene therapy. *Proc. Natl. Acad. Sci. USA* **99**, 10405–10410.

Xie, Q., Ongley, H. M., Hare, J., and Chapman, M. S. (2008). Crystallization and preliminary X-ray structural studies of adeno-associated virus serotype 6. *Acta Crystallogr.* **64**, 1074–1078.

Yan, Z., *et al.* (2002). Ubiquitination of both adeno-associated virus type 2 and 5 capsid proteins affects the transduction efficiency of recombinant vectors. *J. Virol.* **76**, 2043–2053.

Yang, Q., *et al.* (1998). Development of novel cell surface CD34-targeted recombinant adenoassociated virus vectors for gene therapy. *Hum. Gene Ther.* **9**, 1929–1937.

Yang, L., *et al.* (2009). A myocardium tropic adeno-associated virus (AAV) evolved by DNA shuffling and *in vivo* selection. *Proc. Natl. Acad. Sci. USA* **106**, 3946–3951.

Yao, V. J., *et al.* (2005). Targeting pancreatic islets with phage display assisted by laser pressure catapult microdissection. *Am. J. Pathol.* **166**, 625–636.

Yu, C. Y., *et al.* (2009). A muscle-targeting peptide displayed on AAV2 improves muscle tropism on systemic delivery. *Gene Ther.* **16**, 953–962.

Zhong, L., *et al.* (2004). Impaired nuclear transport and uncoating limit recombinant adeno-associated virus 2 vector-mediated transduction of primary murine hematopoietic cells. *Hum. Gene Ther.* **15**, 1207–1218.

Zhong, L., et al. (2008). Next generation of adeno-associated virus 2 vectors: Point mutations in tyrosines lead to high-efficiency transduction at lower doses. Proc. Natl. Acad. Sci. USA 105, 7827–7832.

Zhou, X., and Muzyczka, N. (1998). In vitro packaging of adeno-associated virus DNA. J. Virol. 72, 3241–3247.

Zincarelli, C., Soltys, S., Rengo, G., and Rabinowitz, J. E. (2008). Analysis of AAV serotypes 1–9 mediated gene expression and tropism in mice after systemic injection. Mol. Ther. 16, 1073–1080.

Zolotukhin, S. (2005). Production of recombinant adeno-associated virus vectors. Hum. Gene Ther. 16, 551–557.

Zolotukhin, S., et al. (1999). Recombinant adeno-associated virus purification using novel methods improves infectious titer and yield. Gene Ther. 6, 973–985.

3

Tissue-Specific Targeting Based on Markers Expressed Outside Endothelial Cells

Mikhail G. Kolonin
The Brown Foundation Institute of Molecular Medicine for the Prevention of Human Disease, The University of Texas Health Science Center at Houston, Houston, Texas 77030, USA

I. Introduction
    A. Differentially expressed cell surface receptors as tissue markers and drug targets
    B. Organ-specific endothelial targets
    C. Organization of the vasculature and nonendothelial tissue access
II. Nonendothelial Tissue-Specific Targets of Systemically Administered Agents
    A. Extracellular matrix markers
    B. Perivascular cell markers
    C. Interstitial stromal cell markers
    D. Parenchymal cell markers
    E. Hematopoietic cell markers
    F. Stem cell markers
III. Concluding Remarks and Future Directions
    Acknowledgment
    References

ABSTRACT

Effective management of various diseases could greatly benefit from improved approaches to selective delivery of imaging agents and drugs to the pathological site while sparing other organs. Vascular targeting based on markers expressed on

Advances in Genetics, Vol. 67
Copyright 2009, Elsevier Inc. All rights reserved.
0065-2660/09 $35.00
DOI: 10.1016/S0065-2660(09)67003-6

endothelial cells is the basis for the success of a number of clinical applications. However, targeted treatment of endothelial cells has turned out to often be insufficient for sustained beneficial effects, and the ability to direct therapy to other cells in the organ/tissue may often be essential for long-term efficacy. Therefore, molecular markers associated with the tissue compartments other than the endothelium may represent viable targets for therapeutic, as well as diagnostic, agents. Accumulating evidence indicates that, in both normal development and pathology, permeability of the vasculature may allow an intravenously administered agent to reach cells underlying the endothelium and the interstitial extracellular matrix (ECM). In such cases, perivascular, stromal, and parenchymal cells become exposed to the agent, and can interact with it through differentially expressed cell surface receptors. Infiltrating cells of the immune system may also contribute to the tissue targeting capacity. Markers expressed on these nonendothelial targets accessible to systemically circulating agents may be expressed in an organ-specific manner. Some proteins have been identified to mark both endothelial and nonendothelial cells in the same tissue, which may contribute to the capacity of the agents targeting this tissue to accumulate and be retained in it. The clinical relevance of these markers of deep tissue layers in disease-specific contexts remains to be established. In this chapter, the progress in the understanding of targetable tissue markers expressed outside the endothelium is overviewed. The challenges in the identification and characterization of differentially expressed nonendothelial proteins exposed to circulation are discussed. © 2009, Elsevier Inc.

I. INTRODUCTION

Strategies to timely detect and cure certain cancers and many other diseases have remained ineffective despite the progress in understanding of the underlying molecular events. The methodologies to management of human pathologies have relied largely on untargeted agents that have shown insufficient sensitivity or led to side effects in vital organs (Moghimi et al., 2005). Thus, there is a demand for the development of novel-targeted approaches that could direct the delivery of therapeutic and diagnostic agents and improve prevention, detection, and treatment of various diseases. Biodistribution and the ability of an agent to reach different tissues in the body predominantly depends on the nature of the agent, whether it is a drug or a vector used for the delivery of imaging agents or genes. The setback in medicine has been that arming an agent with the functionality necessary to negotiate past numerous lines of defense in the body (biobarriers) in order to reach the target at an effective dose has remained challenging for many types of vectors. In particular, this relates to relatively large delivery vectors, such as nanoparticles. Moreover, many delivery vehicles, such as viruses, often display natural tropisms based on their coat proteins that result in their homing to organs other than those needing treatment. Recent studies show that biodistribution of administered vectors and their directed enrichment can be controlled based on the microanatomical organization and molecular composition at the site of the disease. The comparative effectiveness of antibody derivatives such as Herceptin and Avastin, selectivity of which for tumors is based on the molecular abnormalities at the cancer site, has encouraged research and development of new advanced targeted agents designed to limit the contact of therapeutics with normal tissues.

A. Differentially expressed cell surface receptors as tissue markers and drug targets

Site-directed delivery of therapeutic and imaging agents is based on molecular heterogeneity of tissues exposed to systemic circulation. The illustration to the phenomenon of the organ-specific molecular recognition entities is the natural capacity of various types of cells to actively migrate selectively to individual tissues. In that regard, the best studied are the cells of the hematopoietic system. The homing of hematopoietic stem cells (HSC) to the bone marrow has a defined molecular mechanism, which is the basis for HSC transplantation (Chute, 2006). Chemokines and chemokine receptors (CCRs) represent a class of ligand/receptor systems responsible for cell attraction, whereas the differentially expressed extracellular molecules are responsible for docking of the homing cells. For example, HSC expressing the chemokine receptor CXCR4 home to the bone marrow in response to stromal cell-derived factor-1 (SDF-1), a chemokine secreted by the bone marrow stroma (Mohle et al., 1999). In this case, binding of hematopoietic progenitors appear to occur via E-selectin ligand (ESL-1) and P-selectin glycoprotein ligand (PSGL-1), the adhesion molecules that bind to E-selectin expressed by bone marrow endothelial cells (Voermans et al., 2001).

The chemokine gradients explain why naive lymphocytes home to secondary lymphoid organs, whereas memory/effector lymphocytes home to peripheral organs such as skin and mucosa (Burastero et al., 1998; Fabbri et al., 1999). Mechanisms similar to those controlling HSC migration underlie the recruitment of lymphoid and myeloid cells to the sites of inflammation (Kubes and Kerfoot, 2001; Middleton et al., 2002). Tissue-specific homing receptors mediate leukocyte adhesion to the endothelium activated with the inflammatory signals (Chin et al., 1991). The established molecules involved in this process include intracellular adhesion molecule-1 (ICAM-1), P-selectin, and a number of integrins including α4 (Biedermann, 2001; Hynes, 2002; Sleeman et al., 2001). These receptors appear to bind ligands selectively expressed by subclasses of lymphocytes: CD-18, CD15-E, and vascular cell adhesion molecule-1 (VCAM-1), respectively (Biedermann, 2001; Sleeman et al., 2001).

Organ-selectivity of cancer metastases is another line of evidence for tissue-specific molecules accessible to circulation. It has been proposed that the SDF-1–CXCR4 axis has a pivotal role in both: trafficking of stem cells and metastasis of cells disseminated by certain cancers (Kucia et al., 2005). Engraftment of cancer cells at the secondary site involves interactions within complex networks of cell adhesion and extracellular matrix (ECM) molecules (Edlund et al., 2004). Lectins have been proposed to mediate the docking of the tumor cells to the endothelium, whereas molecules such as integrins—for their subsequent locking in the blood vessel, which is essential for the metastasis take (Honn and Tang, 1992; Scott et al., 2001).

As a result of realization that certain molecules are exposed to circulation only in some organs, but not throughout the body, the concept of vascular targeting has arisen. Differential expression of extracellular proteins in both blood and lymphatic systems has been uncovered through a number of independent biochemical and genetic approaches. These proteins are often not necessarily expressed exclusively in the target tissue, but are rather overexpressed or are modified and/or associate with other molecules in a tissue-selective manner. Nevertheless, such tissue selectivity of their exposure has in many cases been sufficient for targeted agent delivery. Importantly, it has often been possible to capitalize upon vascular heterogeneity without identifying the actual molecules responsible for it. For example, chemical, oligonucleotide, peptide, and antibody libraries have been screened for molecules that home to certain sites in the body (Sergeeva et al., 2006). Enrichment of an individual library derivative in an organ of interest corresponds to a molecular interaction between that derivative acting as a ligand for a tissue-specific receptor. Regardless of whether the receptor identity is known, the ligand can be used for directed delivery of treatment or diagnostic probes to the target site. This concept has been tested and in many instances proved effective in animal models. Establishment of the target receptor identity has followed in some cases based on the library derivatives used as "baits." These initiatives have revealed that, as a rule, tissue-homing peptides mimic the native biological ligands of the tissue-specific receptors. At this point, there is no question that tissue-specific ligand–receptor pairs do exist and can be employed for real clinical applications. In particular, targeting tumor vasculature as a potential approach to cancer treatment has been extensively investigated (Liu and Deisseroth, 2006) and is discussed elsewhere in this book, as well as targeting of other diseases.

B. Organ-specific endothelial targets

The endothelial lumen is the cellular layer primarily exposed to the circulation. It, therefore, can be expected that the majority of tissue-specific vascular targets are represented by endothelial cell surface molecules. A number of receptors differentially expressed in the endothelium in a tissue-specific manner have been identified. One of the earlier observations supporting the concept was that the expression of endothelial leukocyte adhesion molecule-1 (ELAM-1) became induced during inflammation (Bevilacqua et al., 1989). However, the majority of the differentially expressed endothelial receptors have been identified in tumor vasculature. A big part of the success with tumor vasculature-targeted therapies results from the fact that endothelium of even fully formed tumor vasculature has molecular composition different than in other tissues (Kolonin et al., 2001; Pasqualini and Arap, 2002; Pasqualini and Ruoslahti, 1996; Ruoslahti, 2002a). The angiogenic process ongoing in cancer accounts for increased local concentration of tyrosine kinases: Flt-1 and Flk-1 (the receptors for the vascular endothelial growth factor VEGF), along with the coreceptor Neuropilin 1, EPH family members (receptors for ephrins), as well as Tie-1 and Tie-2 receptor for angiopoietin (Brekken and Thorpe, 2001). ECM-binding integrins, such as $\alpha v\beta 3$ and $\alpha v\beta 5$ and adhesion molecules such as E-selectin and endoglin, a coreceptor for the transforming growth factor (TGF-β), are also higher on endothelial cells in tumors, as compared with normal organs. Other miscellaneous molecules, such as annexin A1 and tissue factor (TF), have also been reported as markers of tumor endothelium. Aging-related changes in proteins expressed in endothelial cells have similarly been characterized (Lee et al., 2006).

The molecular phenotyping of cells forming blood vessels at the protein interaction level has been advanced by a number of complementary methodologies (Durr et al., 2004; McIntosh et al., 2002; Valadon et al., 2006; Zhang et al., 2005). Serial analysis of gene expression (SAGE) from microdissected blood vessels from primary colon tumors led to the discovery of several tumor endothelial markers (TEMs). A number of markers of tumor vascular endothelium have been uncovered through bioinformatic analysis of microarray and sequencing data indentifying such proteins as Robo 4 and Delta 4. Interestingly, the molecule found on the surface of a certain cell type is often actually not expressed by the cells carrying it is a marker. Instead, it may be expressed by adjacent cells and, upon secretion, bind to the surface of the cell type that could be targeted with an agent recognizing this molecule as a marker. Indeed, gene expression profile comparisons indicate that molecules differentially expressed between vasculatures derived from tumor tissues and corresponding normal tissues are not cell surface molecules per se, but rather are secreted proteins overexpressed in the vicinity of the target cell. To complicate matters further, some of the bona fide vascular endothelial markers, such as (VE-)cadherin have been identified on the surface of nonendothelial cells such as the glia (Boda-Heggemann et al., 2009). These observations emphasize the notion that gene expression studies often miss the microanatomical contexts of differential molecule exposure. In that respect, proteomics provides more power in identification of functional targets (Liotta *et al.*, 2003). Proteomic approaches based on two-dimensional electrophoresis and liquid chromatography associated with mass spectrometry have been extensively used for identification of differentially expressed vascular proteins (Engwegen *et al.*, 2006; Griffin and Schnitzer, 2008; Gundry *et al.*, 2008; Lee *et al.*, 2006). Vascular genomics and proteomics have critically depended on the bioinformatics-assisted database analysis tools that have evolved in parallel.

Much of the progress in identification of functional endothelial markers has been made possible with the help of the technique called "*in vivo* phage display" (Kolonin *et al.*, 2001; Pasqualini and Ruoslahti, 1996). This methodology enables selection of systemically administered combinatorial peptides displayed on the pIII protein of an M13-derived bacteriophage library based on their homing to specific tissues through targeting receptors selectively expressed in the context of differentially expressed cell surface proteomes (Pasqualini *et al.*, 2001; Sergeeva *et al.*, 2006). An attractive feature of *in vivo* phage display is that it detects the availability of cell surface targets based on accessibility to a circulating probe without preconceived bias about the nature of its corresponding receptor. Random peptide libraries have been screened by us and others for ligands selectively homing to various murine organs including brain and kidney (Pasqualini and Ruoslahti, 1996), lung, skin, pancreas, intestine, uterus, adrenal gland and retina (Kolonin *et al.*, 2006b; Rajotte *et al.*, 1998), muscle (Kolonin *et al.*, 2006b; Samoylova and Smith, 1999), prostate (Arap *et al.*, 2002a), breast (Essler and Ruoslahti, 2002), placenta (Kolonin *et al.*, 2002), and white adipose tissue (Kolonin *et al.*, 2004). Based on the circulation exposure of cell surface receptors, cancer-specific markers have been uncovered for tumor vasculature (Arap *et al.*, 1998; Kolonin *et al.*, 2001; Pasqualini and Arap, 2002; Pasqualini and Ruoslahti, 1996; Ruoslahti, 2002a,b; Sergeeva *et al.*, 2006). The progression of cancer from benign to metastatic state has also been explored using this approach in a mouse pancreatic cancer model (Hoffman *et al.*, 2003; Joyce *et al.*, 2003).

Selectively expressed tumor-specific receptors can be used for therapeutic targeting because they are readily accessible through circulation and often can mediate internalization of ligands by cells (Kolonin *et al.*, 2001; Pasqualini and Arap, 2002; Ruoslahti, 2002a). Because coupling of tissue-homing compounds to a drug limits the systemic exposure of other tissues to untoward effects, target-directed therapeutic approaches may lead to more effective and less toxic disease treatments. Experimental targeting of the receptors overexpressed in the tumor blood vessels has shown reproducible success in animal models (Arap *et al.*, 1998, 2004; Curnis *et al.*, 2002; Ellerby *et al.*, 1999; Hajitou *et al.*, 2006; Koivunen *et al.*, 1999; Kolonin *et al.*, 2004; Marchió *et al.*, 2004). The molecules selected based on their homing to tumors have been used as carriers to guide the delivery of cytotoxic drugs (Arap *et al.*, 1998), proapoptotic and cytotoxic peptides (Ellerby *et al.*, 1999, 2003), metalloproteinase inhibitors

(Koivunen *et al.*, 1999), cytokines (Curnis *et al.*, 2000), imaging agents (Hajitou *et al.*, 2006; Hong and Clayman, 2000), and genes (Hajitou *et al.*, 2006; Trepel *et al.*, 2000) in transgenic and xenograft mouse models of human disease.

The unique architecture of blood vessels undergoing reconstruction provides the basis for directed treatment of not only actively growing but even high-grade tumors with antiangiogenesis agents (Ferrara and Kerbel, 2005). This has suggested angiogenic vasculature as a potentially advantageous therapeutic target not only in cancer but also in other settings (Folkman, 2007). Indeed, regulation of growing adipose and hepatic tissue mass by angiogenesis has been reported (Greene *et al.*, 2003). It has been realized that, like tumors, white adipose tissue could be targeted with antiendothelial agents in order to control tissue expansion. Although white fat is a nonmalignant tissue, cells composing it have the capacity to quickly proliferate and grow (Cinti, 2000; Hausman *et al.*, 2001; Sumi *et al.*, 2001; Wasserman, 1965). Histological evaluation of adipose tissue reveals that fat is highly vascularized: multiple capillaries make contacts with every adipocyte, suggesting the importance of blood vessels for maintenance of the tissue mass (Crandall *et al.*, 1997; Sumi *et al.*, 2001). This has pointed out that white adipose tissue, much like tumors, relies on angiogenesis for adipose mass expansion (Hausman and Richardson, 2004; Voros *et al.*, 2005). We proposed that if antivascular agents could in a similar fashion be directed to mature adipose vasculature, obesity could be not only prevented but also reversed. In our proof-of-principle study aiming to identify and use adipose vascular targets (Kolonin *et al.*, 2004), we selected a combinatorial library in mice for peptides systemically homing to white fat tissue. As a result, we isolated a peptide with the sequence CKGGRAKDC, which homes to and is internalized by adipose endothelium. We biochemically isolated the vascular receptor of CKGGRAKDC from membrane protein extract of adipose cells, identified it as the prohibitin protein, and used it as a target of experimental obesity therapy.

Limitations of animal models in recapitulating human disease may partially account for the relative lack of progress in the clinical translation of targeted therapies. To initiate identification of differentially expressed human vascular targets, we have established a minimally invasive procedure to screen phage display random peptide libraries in live human patients (Arap *et al.*, 2002b; Kolonin *et al.*, 2003). Previous selections of combinatorial peptide libraries had demonstrated that peptides often home to tissue-specific receptors via mimicking binding domains within the corresponding native biological ligands of these receptors; often such protein mimicry occurs at primary structure level (Arap *et al.*, 2002b; Giordano *et al.*, 2001; Kolonin *et al.*, 2002, 2006a,b). To prove the principle that organ-homing peptides isolated through *in vivo* phage display selections can serve as leads for identification of differentially active human protein interactions, we identified a peptide mimicking interleukin-11 (IL-11) and validated interleukin-11 receptor (IL-11Rα) upregulation in human prostate and metastatic prostate cancer (Arap *et al.*, 2002b; Cardo-Vila *et al.*, 2008; Zurita *et al.*, 2004). Our originally controversial initiative to select peptide libraries in live patients (Arap *et al.*, 2002b) has been followed up by efforts of other research groups (Krag *et al.*, 2006).

C. Organization of the vasculature and nonendothelial tissue access

Various aspects of endothelial vascular targeting are covered elsewhere in the book and will not be discussed in detail here. The focus of this chapter is vascular targets expressed outside of the endothelial (lumen-exposed) cells. Such tissue-specific markers include ECM molecules, as well as receptors accessible on perivascular, stromal, and parenchymal cells (Fig. 3.1). Most of what we currently know about the receptor systems targetable on nonluminal molecules comes from studying cancer biology. While tumors can certainly be outstanding in terms of their vasculature and the organization of individual tissue compartments, similar principles have been shown to apply to a number of normal organs in homeostatic conditions.

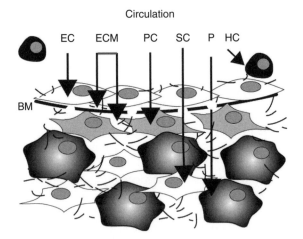

Figure 3.1. Vasculature organization and its compartments as drug targets. Endothelial cells (EC) lining the lumen are exposed to circulation and, therefore, to the administered agent. Certain agents can access and penetrate through fenestrations in the basement membrane (BM) and reach underlying tissues (arrows), including perivascular cells (PC), interstitial stromal cells (SC), and cells of the parenchyma (P), as well as the extracellular matrix (ECM). The accessibility of nonendothelial targets to drugs is favored by certain pathological conditions, such as cancer or injury associated with inflammation. Infiltrating hematopoietic cells (HC), such as monocytes/macrophages and lymphocytes, may also acquire tissue-specific features and potentially serve as drug targets.

What makes it reasonable to expect that nonendothelial targets can be accessible to circulation? For one thing, it is the cell type specificity with which different viruses infect tissues once in the bloodstream, with the endothelium not always being the primary target. For example, our laboratory had previously demonstrated this phenomenon for peptides homing to tumor vasculature via binding to aminopeptidase-N (CD13), abundantly expressed on perivascular cells (Kontoyiannis *et al.*, 2003). Most of the evidence, however, comes from various clinical trials accumulated over the years. Certainly, only a small fraction of the compounds and vectors that have been tested in the clinic had been designed to have an effect on endothelial cells. Instead, parenchymal cells, such as tumor cells, have often been used *in vitro* in drug library screens without specifically selecting for compounds that would gain access to the target *in vivo*. Indeed, against all odds, the drug prototypes selected have often reached the parenchymal cells when administered systemically and turned out to be effective. In some pathological situations there is a biological explanation as to why that has been possible. Vascular endothelium can acquire several phenotypically distinct types: continuous, fenestrated, and discontinuous, thus constituting vessels that are distinct in their permeability properties. These morphological and physiological differences are thought to be controlled by cell–matrix interactions and are attenuated by pathological signals. In the following section, we are discussing a number of molecules exposed to circulation on nonendothelial cells in specific organ-dependent contexts.

II. NONENDOTHELIAL TISSUE-SPECIFIC TARGETS OF SYSTEMICALLY ADMINISTERED AGENTS

Cell surface molecules on nonendothelial cells can be exposed in both normal physiological and pathological conditions depending on the tissue of interest. The bone marrow is an example of an organ featuring marked fenestration of the microcapillaries that allow direct exposure of blood to the underlying stroma and differentiating hematopoietic cells. It has been long known that some human cancers—such as prostate and breast cancer—have marked selectivity for bone metastasis (Fidler, 1997). Various bone marrow cell-secreted molecules have been tested for interactions that could mediate homing of metastatic tumor cells to the skeleton (Choong, 2003; Honn and Tang, 1992; Keller *et al.*, 2001; Reddi *et al.*, 2003). Endothelial permeability has also been reported for other normal organs, where it can increase in certain physiological settings. Tissue injury, associated with inflammation, is another pathological condition in which organ integrity is disrupted and vascular permeability is enhanced. In addition to expression of extracellular molecules on organ-resident cells, drug target tissue specificity can be contributed to by infiltrating cells of the immune system.

While only a limited number of lymphocyte subtypes and macrophage polarization variants are currently established, the complexity of the immune system differentiation is far from being completely understood. Therefore, it is reasonable to expect that the infiltrating leukocytes may contribute to the selectivity of the molecular repertoire accessible to the circulation at the sites of their recruitment and extravasation. The molecules presented by these different types of cells could potentially be used for directing imaging or therapy agents to the corresponding organs.

Cancer is one of the extreme pathological settings where circulating molecules most definitely can engage in contact with the surface of cells other than the endothelium. Tumor vasculature can vary dramatically in terms of organization in different cancer types. Tumor angiogenic vessels tend to be disorganized, contain blind end loops, and are often hypoxic. They usually have abnormal pericyte coverage and relatively little muscle coverage. As a result, tumor vessels can become leaky upon the action of vascular permeability factors such as VEGF. Indeed, such leakage is contributed by the fenestration of the endothelial layer, which is often observed. In addition, the integrity of the perivascular lining is often distorted in tumors, thus enabling blood contact with the parenchymal cells directly. Moreover, there are certain types of cancer in which the lumen is actually composed of nonendothelial cells. Endothelial and pericyte abnormalities observed in cancer suggest that endothelial mimicry plays an important role in tumor vascularization (McDonald and Foss, 2000). More recently, vascular permeability has also been shown to predetermine access of circulating molecules to nonluminal cells in benign tissues (Bazan-Peregrino et al., 2007; Liu and Deisseroth, 2006).

The expression of parenchymal cell surface antigens is known to be influenced by the organ-specific ECM/connective tissue context, and vice versa (Mackenzie and Dabelsteen, 1987). As a rule, rather than as an exception, molecules that have been identified as luminal vascular markers are expressed not exclusively in the endothelium but also on other cells of the same organ. Examples include IL-11Rα (Zurita et al., 2004), prohibitin and certain annexins (Kolonin, unpublished), caveolin, Sca-1, and many other proteins previously tested as prospective targets. This phenomenon is likely to have an important physiological significance. For some receptor systems, expression on different cell types, such as (i) the endothelium and the pericytes or (ii) the endothelium and the parenchymal cells relates to their implication in molecular transport. For instance, nutrients from the bloodstream, which first interact with the lumen, are destined to translocate to the underlying parenchyma, which is achieved by active transport. Expression of the transporters on cells adjacent to the endothelium may enable the passing of the nutrient through the subsequent cell layers, with endocytosis mediated by the same cell system on each cell type. This model of transcytosis makes biological sense because it minimizes redundancy that would be required if different transporters were to regulate the molecular uptake on vascular, perivascular, and the deeper cell layers. Similar considerations could apply to the transporter systems mediating the excretion of molecules from the parenchymal cells into the systemic circulation. This scenario is relevant in cases such as white adipose tissue, the skeletal muscle, or other organs whose function is to actively uptake nutrients for energy storage, or to exocytose metabolites during the release of the stored energy.

An important consideration is the cell type-selective surface binding of soluble molecules/ligands of the receptors that are not expressed by the target cells exclusively, but that rather attract their ligands locally present at a concentration higher than elsewhere in the body. One protein that displays differential deposition on the cell surface in individual tissues is SPARC (secreted protein, acidic and rich in cysteine). SPARC, originally discovered as osteonectin (BM-40) abundant in bone (Termine et al., 1981), is a matricellular protein that modulates interactions between cell surface and ECM molecules, thus controlling cell adhesion, migration, survival, and differentiation (Framson and Sage, 2004). Highly conserved among mammals, SPARC is the prototypic protein for a family grouped on the basis of a unique, extracellular Ca^{2+}-binding module, a self-folding domain that is preceded by a follistatin-like module (Bornstein and Sage, 2002). Recent studies have suggested that, due to its local overexpression, SPARC may mediate targeting of agents to cells of white adipose tissue and tumors (Framson and Sage, 2004; Nie et al., 2008). Another example of this type of tissue markers is proteinase 3 (myeloblastin), which is expressed by myeloid cells but is deposited onto the surface of the endothelial cells, as well as into the ECM. Our unpublished studies have uncovered PR3 as a receptor for molecules involved in bone marrow homing. The S100 proteins (Roth et al., 2003) may also have such properties. The 16 members of this family are involved in many extracellular processes through interacting with specific target proteins and show a divergent pattern of cell-specific expression, consistent with their pleiotropic extracellular functions and appear to be bound to the endothelial surface in tissue-specific contexts. These and other soluble factors may be useful by serving as vehicles bound by the therapeutic or diagnostic agent and delivering the agent to the target cell via binding to their receptor on the cell surface. In certain cases, active transport mechanisms may make these molecules useful for delivering treatment/imaging agents to cells that are not directly exposed to the blood.

In the following sections, a number of proteins that could serve as nonendothelial tissue-selective targets are discussed (Table 3.1). Some of these markers have been previously studied in experimental targeting applications, other are still yet to be recognized. No list of this kind can be comprehensive enough. New marker candidates emerge all the time, whereas the proteins considered as targets to this point are often discarded in continuous preclinical and clinical tests due to uncovered off-target bioavailability. Therefore, the

Table 3.1. Examples of Proteins Potentially Useful as Targetable Markers of Nonendothelial Cells

Marker/expression	Endothelial	ECM	Perivascular	Interstitial stromal	Parenchymal	Infiltrating hematopoietic
Collagens	+	++	+/-	+/-	+/-	+/-
Fbn	+	++	+/-	+/-	+/-	+/-
MMPs	+	++	+/-	+/-	+/-	+/-
SPARC	+	++	+	+/-	+/-	+/-
PDGFR	-	-	++	+/-	-	+/-
CD13	+	-	++	+/-	-	+/-
NG2	+/-	-	++	+/-	-	+/-
FAP	-	-	+	++	-	-
Cadherin-9	-	-	-	++	-	+/-
EGFR	+	-	+	+	++	+/-
ErbB2	+	-	+	+	++	+/-
Eph receptors	+	-	+	+	++	+/-
IL-11Rα	+	-	+	+	++	+/-
GRP78	-	-	+/-	+/-	++	+/-
CRKL	-	+	+/-	+/-	++	+/-
PRLR	+	-	+/-	+/-	++	+/-
CD163	-	-	+	-	+/-	++
CCRs	+/-	-	+/-	+/-	+/-	++

++ indicates molecules potentially useful as markers of the corresponding cell types or the ECM.

primary role of this chapter is not to catalog tissue-specific proteins, but is rather to discuss the considerations that need to be taken in pursuing nonendothelial markers for diagnostic and therapeutic targeting purposes.

A. Extracellular matrix markers

Molecules composing the ECM are located on both sides of the endothelial layer, the luminal and the basal, the latter including the basement membrane underlying the endothelium in most types of vessels. Many of the ECM proteins are secreted by endothelial cells and, therefore, could formally still be considered endothelial markers. ECM proteins are, therefore, obvious as potential markers of organs' vasculatures. There is plenty of experimental evidence that ECM surrounding the endothelium can present organ-specific molecular contexts that are modulated by different physiological and pathological situations. It has been proposed, and in some cases proved, that cell surface molecules or ECM within the bone marrow could serve as tissue-specific homing receptors for prometastatic tumor cells (Choong, 2003; Reddi et al., 2003). As a result, a number of ECM molecules have been invoked to explain the site-specificity of skeletal metastasis, some of which have been confirmed as potential tissue-specific targets.

The two broad categories of ECM in tissues are basement membranes and interstitial/stromal matrix (Bissell and Radisky, 2001). Extracellular basement membrane is the ECM component, to which basal surfaces of endothelial cells attach (Plopper, 2007). The basement membrane coordinates spatial and molecular signals that influence endothelial cell proliferation, migration, and differentiation (Hallmann et al., 2005). Because the basement membrane sleeve around the endothelial tube is the first structure after the endothelium that is exposed to circulation, it plays an important role in the maintenance of vessel wall integrity. In pathologies, such as cancer, basement membranes in blood vessels are morphologically abnormal (Baluk et al., 2003), suggesting a corresponding change in their molecular composition. Vascular basement membranes are composed of type IV collagens, laminins, heparan sulfate proteoglycans, and nidogens. Importantly, isoforms of all four classes of molecules have been identified, which can assemble into basement membranes with distinct structure and functions. Laminins are composed of 5α, 4β, and 3γ chain glycoproteins that form 15 heterotrimer combinations. Laminin isoform composition appears to be particularly important in respect to basement membrane tissue specificity. The laminin alpha-chains are particularly key as far as tissue selectivity, as they exhibit tissue-specific distribution patterns and contain the primary cell interaction sites. Laminin isoform representation in the vasculature varies depending on the developmental stage, vessel type, and the activation state of the endothelium. The two predominant vascular heterotrimers are Laminin 8 and Laminin 10. Laminin 8, composed of laminin $\alpha4$, $\beta1$, and $\gamma1$ chains, is relatively ubiquitously expressed and is upregulated by inflammatory cytokines and growth factors. Laminin 10, composed of laminin $\alpha5$, $\beta1$, and $\gamma1$ chains, is expressed primarily in postnatal endothelial cell basement membranes of capillaries and venules in response to strong proinflammatory signals and angiostatic agents such as progesterone. Other ECM molecules, such as thrombospondins 1 and 2, fibronectin (Fbn), nidogens 1 and 2, and collagen types VIII, XV, and XVIII, are also differentially expressed by different vasculatures, varying with the endothelium type and/or pathophysiological state. For example, Fbn has been revealed to account for ECM tissue specificity by being expressed as at least 20 alternative splicing isoforms (White et al., 2008).

Interstitial ECM composition has been exploited for directed delivery of agents of interest. Cancer is a pathological condition, in which alterations in ECM composition is perhaps the most documented, although similar sets of proteins have been identified to mark various fibrotic pathologies such as idiopathic pulmonary fibrosis (IPF). In addition to neovascularization and infiltration of inflammatory cells, cancer progression is accompanied by fibrosis, which is hallmarked by activation of myofibroblasts and ECM remodeling resulting in excessive collagen deposition. This desmoplastic response relies on mechanisms governing tissue injury and repair, which are chronically activated in cancer (Dvorak, 1986). Various collagens, fibronectin, matricellular proteins, and matrix metalloproteinases (MMPs) have been uncovered as ECM targets associated with cancer progression and wound repair (van Beijnum et al., 2009). Selective peptide inhibitors of gelatinases MMP-2 and MMP-9 in xenograft models localized to murine tumors and prevented tumor growth upon systemic administration (Koivunen et al., 1999). These findings are consistent with the upregulation of these metalloproteases during tumor invasion and angiogenesis. Interactions among matrix molecules has made it possible to use the ECM proteins themselves for generation of ECM-binding moieties. Targeting of the immunoglobulin-binding domain of Protein A by using a Fbn fragment has been reported (Matsuyama et al., 1994). A polymeric form of Fbn has been shown to prevent tumor cell engagement with the ECM through blocking integrin binding sites, which resulted in a dramatic reduction of the aggressiveness of experimental tumors (Pasqualini et al., 1996).

A number of other ECM molecules have recently been uncovered as previously unrecognized tissue-specific markers. Fibrillin-1, a large ECM glycoprotein that assembles to form microfibrils, displays tissue-specificity of expression, which accounts for site-specific localization of BMP-4, a growth factor binding to fibrillin-1 (Sengle et al., 2008). Asporin is another ECM protein found primarily in regions surrounding skeletal tissue and is upregulated in diseases including osteoarthritis, rheumatoid arthritis, and lumbar disk disease. Asporin belongs to the small leucine-rich repeat proteoglycan (SLRP) family, interacts with growth factors such as TGF-β and BMP-2, and represents a promising target for pharmacogenomic approaches to treatment of bone and joint diseases (Ikegawa, 2008).

A separate group of extracellular molecules forth discussing encompasses matricellular proteins that modulate interactions between cell surface and ECM molecules, thus controlling cell adhesion, migration, survival, and differentiation. We recently reported SPARC as a previously unknown ligand of β1 integrin (CD29), an adhesion molecule marking stromal cells. The diverse functions of SPARC include mediation of cell interactions with integral ECM components, Fbn deposition, and actin stress fiber reorganization leading to cell deadhesion (Bornstein and Sage, 2002; Framson and Sage, 2004). Our studies revealed binding of SPARC to the α5β1 integrin complex at focal adhesions, the cell surface linkages to the ECM known to depend on integrin interaction with Fbn (Nie et al., 2008). In addition to cancer cells, SPARC is abundantly secreted by the endothelium and adipocytes (Framson and Sage, 2004), which accounts for its role in adipose tissue (Bradshaw et al., 2003).

Because matrix molecules, when bound by organ-specific probes, do not come in contact with endothelial cell surface and are unlikely to be internalized by cells, their applicability is limited, as compared with cell surface receptors. While still useful for imaging applications, ECM molecules are in general not ideal as receptors for cytotoxic therapeutics or gene therapy vectors. In that respect, surface receptors of endothelial and nonendothelial cells have a better potential to be clinically useful as vector targets.

B. Perivascular cell markers

Perivascular cells surrounding the blood vessel are located to the other side of the basement membrane underlying the endothelium. These perivascular cells in smaller blood vessels are known as fibroblastic pericytes or mural cells and in larger blood vessels are also joined by smooth muscle cells. The perivascular cell layer has been shown to maintain the integrity and regulate permeability of blood vessels. In addition, pericytes clearly present a paracrine compartment releasing factors essential for survival and proliferation of endothelial cells (Traktuev et al., 2008). We and others have shown that perivascular stromal cells in white adipose tissue and other organs, in addition to maintaining vessels and cooperating with the endothelium during vascularization, display characteristics of progenitor cells (Tang et al., 2008; Traktuev et al., 2008). Accumulating evidence indicates that pericytes arise from or possibly are represented by mesenchymal stromal cells (MSC) that coincidently serve as multipotent progenitors in various normal organs (Crisan et al., 2008). Originally, MSC have been isolated from the bone marrow stroma and termed fibroblast colony-forming units (CFU-F) based on their fibroblastic morphology (Friedenstein, 1980). The ability of MSC to differentiate into tissues of mesenchymal origin, such as bone, cartilage, and adipose tissue, has resulted in MSC commonly being referred to as mesenchymal stem cells (Pittenger *et al.*, 1999; Prockop, 1997).

Changes in perivascular call layer in tumor vessels are now well documented. For example, pericytes express desmin but not α-smooth muscle actin (αSMA) on capillaries in normal pancreatic islets, whereas they express both desmin and αSMA in mouse RIP-Tag2 insulinoma model. In this and other tumor models, pericytes feature multiple phenotypical abnormalities, suggesting altered expression of other marker proteins (Morikawa *et al.*, 2002). Because PDGF receptors are overexpressed in tumor perivascular cells, it has been suggested that pericytes in tumors present a complimentary target to endothelial cells, which could be advantageous for antiangiogenic therapy (Bergers *et al.*, 2003). Using a PDGF receptor (PDGFR) inhibitor STI571 in combination with VEGFR inhibitor AEE788 (targeting the endothelium) improved experimental ovarian carcinoma therapy efficacy (Lu *et al.*, 2007). Moreover, PDGFR-mediated disruption of pericyte support of tumor endothelial cells with tyrosine kinase inhibitors Imatinib and SU11248 gave complete responses and unprecedented survival advantage in mouse models of cancer when combined with sequential MTD and then metronomic chemotherapy and antiangiogenic therapy (Pietras and Hanahan, 2005). Interestingly, antiangiogenic therapy alone, without cotargeting of nonendothelial tumor compartments may select for more aggressive tumor variants (Paez-Ribes *et al.*, 2009), which emphasizes the clinical relevance of the perivascular targets.

Various studies focused on other pericyte markers have unequivocally confirmed the importance of these cells as a potential clinically important target in cancer. Aminopeptidase N (CD13) is an established marker of perivascular cells. It has been identified as a receptor for tumor-homing peptides containing the NGR motif that has been used to deliver cytotoxic drugs to cancer in preclinical models (Arap *et al.*, 1998; Curnis *et al.*, 2002; Pasqualini *et al.*, 2000). In a similar approach, Burg *et al.* isolated peptides that bind *in vitro* to NG2, a pericyte-specific proteoglycan selectively expressed in angiogenic vasculature, and demonstrated homing of these peptides to mouse tumors. Targeting to NG2 was elegantly confirmed by the absence of tumor homing of these peptides in an NG2 knockout mouse tumor model (Burg *et al.*, 1999). In addition, therapeutic interference with NG2-neutralizing antibody has been shown to decrease prostate cancer neovascularization and control tumor growth (Ozerdem, 2006).

A number of other markers have been found overexpressed in pericytes in certain contexts. Consistent with being one of the commonly used MSC markers, Thy-1 (CD90) labels the pericyte sheath surrounding the postcapillary venules and connective tissue elements in lymphoid tissues (Barclay, 1981). Endosialin (CD248), initially isolated as a tumor endothelial cell marker (TEM1), has been found expressed on pericytes in the central nervous system and in high-grade gliomas (Christian et al., 2008). Perivascular tumor cells also express high levels the SDF-1, as well as proteins involved in lactate absorption (MCT1/MCT2) and lactate oxidation (high LDH1 and low HIF/LDH5), which are bound to the cell surface and could potentially serve as molecular targets on these cells (Koukourakis et al., 2006).

Our laboratory has initiated a systematic screen for markers of perivascular stromal cells based on phage display technology. Recently, we reported integrin β1 (CD29) as an adipose stromal cell (ASC) marker bound by peptides that mimic SPARC, thus identifying this protein as a novel biological β1 ligand (Nie et al., 2008). To further advance this initiative, we have continued our efforts to identify markers of circulation-accessible organ-specific MSC by screening peptide libraries in live mice. We have combined our in vivo phage display expertise with the methods optimized for isolation of stromal cells by fluorescence-activated cell sorting (FACS; Short et al., 2009; Simmons and Torok-Storb, 1991), thus establishing a novel integrated platform approach (Fig. 3.2). We screened a mixture of cyclic CX_7C, CX_8C, and CX_9C (C, cysteine; X, any residue) peptide libraries by using the previously described "synchronous" method (Kolonin et al., 2006b).

As a result, we isolated a panel of ligands that, when injected in vivo, home to markers that are expressed differentially on $C34^+CD31^-CD45^-$ (nonendothelial) cells derived from the stromal/vascular fraction of one of the following organs: lung, white adipose tissue, bone, or skeletal muscle. A number of peptide probes have been tested individually and were validated to home to perivascular cells of individual organs (unpublished data). As an example, we are showing homing of phage displaying a lung-homing peptide specifically to lung perivascular cells and homing of phage displaying a white adipose tissue-homing peptide specifically to adipose perivascular cells, as revealed by confocal immunofluorescence with anti-phage and anti-CD31 antibodies (Fig. 3.3). The results of our screen show that phage displaying the homing peptides extravasate through the endothelium and gain access to pericytes when delivered systemically. This suggests that it will also be possible to directly target perivascular cells in vivo.

C. Interstitial stromal cell markers

The cells discussed as potential targets in this section correspond to stromal cells that are not in direct contact with the vascular wall, but are rather intertwined with parenchymal cells in deeper layers of the organ, as opposed to perivascular stromal cells underlying the basement membrane. Such interstitial stromal cells are commonly termed tissue resident fibroblasts, although this term does not really address the endogenous function of these cells that are by far not the only cells acquiring fibroblastic appearance in tissue culture. Indeed, the distinction between different subpopulations of MSC-like fibroblastic cells is poorly

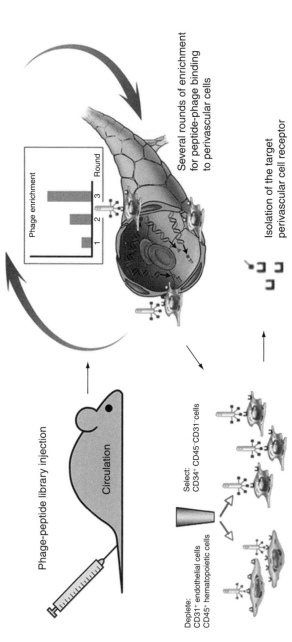

Figure 3.2. The design of a screen for receptors marking perivascular mesenchymal cells in specific organs. Phage-peptide library is injected into an animal and several (3–4) rounds *in vivo* biopanning are performed to enrich for organ-homing peptides. In every selection, round phage are i.v.-administered and recovered from the organ stromal/vascular fraction, amplified, and used for the next round. Increased recovery of phage in each subsequent round reflects the selection of homing peptides. In each round, peptides binding to stromal/vascular fraction in control organs are also isolated to control for specificity of enrichment. FACS is used to deplete endothelial (CD31⁺) and hematopoietic (CD45⁺) cells with the bound phage-peptides and to collect phage-peptides bound to perivascular mesenchymal (CD34⁺CD31⁻CD45⁻) cells. After the binding phage-peptide isolation, the peptide is used as bait to biochemically purify the targeted receptor. (See Page 3 in Color Section at the back of the book.)

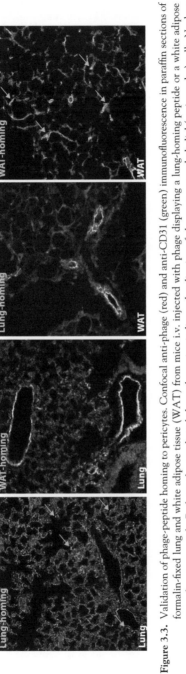

Figure 3.3. Validation of phage-peptide homing to pericytes. Confocal anti-phage (red) and anti-CD31 (green) immunofluorescence in paraffin sections of formalin-fixed lung and white adipose tissue (WAT) from mice i.v. injected with phage displaying a lung-homing peptide or a white adipose tissue-homing peptide. Red signal upon digital channel merging indicates localization of phage on nonendothelial (perivascular) cells. Nuclear TOPRO3 staining is blue. (See Page 4 in Color Section at the back of the book.)

understood due to the lack of markers (Bianco et al., 2008). In reality, factors discussed here have not been specifically shown to be absent in the perivascular compartment and, indeed, could be present in all mesenchymal cells of the organ. Such imprecision in the current discussion relates to the fact that in situ expression pattern has not been investigated in sufficient detail for every particular protein reviewed. In addition, such a distinction is artificial for some organs, such as white adipose tissue and lung. Indeed, in these cases the stromal layers are so thin between the vessels and the functional cells of the organ that it is not practically possible to make a distinguish perivascular and interstitial stroma. Therefore, it should be realized that the considerations of markers discussed here for stromal cells may often be applicable to perivascular cells as well.

One of the molecules potentially useful as a marker of stromal cells is the fibroblast-activation protein-alpha (FAP). FAP is a membrane-bound serine protease that is expressed predominantly on the surface of fibroblasts in a pattern partially overlapping with the αSMA expression pattern. Whereas αSMA is expressed in many normal perivascular and smooth muscle cells, FAP is minimally expressed by normal tissues, thus only marking a subpopulation of cells in the reactive stroma in most epithelial tumors. Recently, FAP-selective peptidic protoxins were constructed through modification of the prodomain of melittin, an amphipathic cytolytic component in the venom of the common European honeybee Apis milefera. Peptide protoxins selectively toxic to FAP-expressing cell lines, when injected intratumorally, induced lysis and growth inhibition of human breast and prostate cancer xenografts with minimal toxicity to the host animal (Lebeau et al., 2009).

There are reported transcriptome profile differences between stromal cells isolated from different organs (De Ugarte et al., 2003; Lee et al., 2004; Noel et al., 2008; Wagner et al., 2005). The distinct promoter CpG methylation signatures unique for stromal cells of individual organs (Boquest et al., 2006) support the possibility that stromal fibroblasts in different organs may also be distinct in terms of their cell surface proteome. Consistent with this possibility, cadherin-9 has been revealed as a cell surface marker of fibroblasts in the kidney (Thedieck et al., 2007). Further supporting this notion, the repertoires of chemokine receptors expressed on the surface of stromal cells derived from different organs have been found only partially overlapping (Chamberlain et al., 2008). A number of other cell surface molecules, which may have value by providing a way to enrich therapeutic or imaging vectors agent delivery to stromal cells, are also worth mentioning. CD10, although also expressed on leukocytes, has been revealed as a marker for normal and neoplastic endometrial stromal cells (McCluggage et al., 2001). Other stromal cell-associated markers, such as CD44, CD29, CD63, CD73, CD90, CD49a, and CD166, are currently being explored as potential targets of stromal cells. Although these molecules are also not specific for stromal cells, they could contribute to tissue accumulation of agents directed to them.

D. Parenchymal cell markers

The organ identity is ultimately defined by the functional parenchymal cells that are of epithelial origin in the majority of internal organs. There are reliable markers established for many types of functional cells, often related to the product that the cells secrete, such as insulin in the pancreatic beta cells. Targeting these functional cells would be straightforward if the cell surface markers were accessible to the bloodstream. However, this probably rarely happens, and the conditions at which circulation can directly access the underlying parenchymal cells are not well understood. It is likely that deep tissue cell access varies organ to organ, depending on the vascular permeability.

Cancers create a unique setting in which tumor cells causing the disease, although separated from blood by the vascular wall, are often accessible by circulation. Cancer progression is caused by and/or associated with consecutive vascular permeability elevation and alterations in protein expression in tumor cells (Engwegen et al., 2006; Hanahan and Weinberg, 2000; Pasqualini and Arap, 2002; Ruoslahti, 2002a; Steeg, 2006; Vogelstein and Kinzler, 2004). The exposure of tumor cells to blood may be further facilitated by hypoxia and the resulting necrosis and hemorrhaging. Numerous tumor cell surface markers have been identified and many of them have been proved accessible to systemically administered agents and therefore subject to directed therapies (Chatterjee and Zetter, 2005; Samoylova et al., 2006). Perhaps, the receptor systems the most well established as clinical targets in solid tumors are EGFR and ErbB2 (Her-2/Neu) family of receptor tyrosine kinases (Maihle et al., 2002; Schmitz and Ferguson, 2009). There are a number of effective cancer drugs available targeting these receptor systems, thus indicating their accessibility. Protooncogene tyrosine kinase receptor c-Kit (CD117) may be another promising target on tumor cell surface, as recently shown with small molecules designed to treat human gastrointestinal tumors (Gunaratnam et al., 2009). Many other cancer markers, although proved to be selective as tumor targets preclinically, are yet to be validated by clinical trials.

A number of studies, including the ligand-directed surface profiling of human cancer cells with combinatorial peptide libraries performed by our group, have revealed differential expression of ephrins and their Eph receptors in both benign human tissues and cancers (Hafner et al., 2004; Kolonin et al., 2006a). Our studies have also uncovered overexpression of IL-11Rα not only in pathological prostate vasculature but also on tumor cells in osteosarcoma and advanced prostate cancer. One of the better established prostate tumor markers is the prostate stem cell antigen (PSCA), a cell surface antigen that belongs to the Ly-6/Thy-1 family of glycosylphosphatidylinositol-anchored proteins. PSCA is highly overexpressed in human prostate cancer, with limited expression in normal tissues (Amara et al., 2001; Raff et al., 2009). Another example is the certain members of S100 calcium-binding protein family, which are secreted and can discriminate between prostate cancer and benign tumors (Hermani *et al.*, 2005). Their receptor RAGE (receptor for advanced glycation end products) appears to also serve as a marker of advanced prostate cancer. Many matricellular proteins, including SPARC, osteopontin, and tenascin, are expressed by tumors in many types of cancer (Clark and Sage, 2008; Framson and Sage, 2004). In patients, overexpression of SPARC predicts poor prognosis of a number of adenocarcinomas and correlates with the invasion and metastatic dissemination (Clark and Sage, 2008; Framson and Sage, 2004). Because, as we have shown, SPARC interacts with the integrin machinery and, therefore, is bound to the tumor cell surface, it may represent a cancer drug target.

An interesting case of tumor cell markers is featured by cancer antigens that are normally intracellular but that may become overexpressed and presented on the cell surface in cancer. While many such proteins, like proteinase-3, mark hematological malignancies (Molldrem *et al.*, 2002), some of them have been found specifically on the surface of cells in solid tumors, including NY-ESO-1, MAGE-A, and SSX (Scanlan *et al.*, 2002). A number of solid tumor targets have been identified by phage display. One of such proteins is the stress response chaperone GRP78, cell surface expression of which enables tumor targeting by circulating ligands (Arap *et al.*, 2004). Another such case is CRKL, an adapter protein with Src homology 2 (SH2)- and SH3-containing domains (Mintz *et al.*, 2009). An unrecognized extracellular function for CRKL released from the cytoplasm into the tumor microenvironment along and with the plexin–semaphorin–integrin domain of β1 integrin illustrates the notion that the previously established subcellular localization may change for many proteins in pathological conditions and generate markers with clinical relevance. It should be noted that, in many cases, peptide probes isolated based on their capacity to directly target tumor cells *in vivo* have remained orphan ligands (Hong and Clayman, 2000). Identification of receptors for these probes is underway and is predicted to result in many new clinical target in the coming years.

While the majority of deep tissue-accessible targets are available for malignancies, biopanning on benign organs has also uncovered interesting markers. For example, we have previously demonstrated that a peptide CRVASVLPC that mimics prolactin receptor (PRLR) ligands, when intravenously administered into mice, homes to the pancreas in a pattern corresponding to the endogenous PRLR expression. Importantly, this peptide was found localized not only to the pancreatic endothelium but also to the pancreatic islet cells upon homing (Kolonin *et al.*, 2006b). Cells of other digestive system organs, and of metabolically active organs in general, express various cell surface transporters of nutrients. It has been found that nutrient transporters often display tissue specificity of expression. As an illustration of this notion, uptake of glucose and of related hexoses through the cell membrane is assisted by integral membrane glucose carriers coded by genes of GLUT/SLC2 (solute carrier) family

(Charron and Kahn, 1990). To date, at least 13 members of the GLUT family have been identified. Each glucose transporter isoform plays a specific role in glucose metabolism determined by its pattern of tissue expression, substrate specificity, transport kinetics, and regulated expression in different physiological conditions. For instance, GLUT4 is the glucose transporter selectively expressed in adipose tissue (in addition to striated muscle) and is primarily responsible for insulin-regulated glucose uptake by adipocytes (Buse *et al.*, 1992; Katz *et al.*, 1995).

As compared with glucose uptake regulation, the molecular control of fatty acid uptake is not as well understood (Berk and Stump, 1999; Su and Abumrad, 2009). Fatty acid transport proteins (FATPs/solute carrier family 27) are integral transmembrane proteins that enhance the uptake of long-chain fatty acids (LCFAs) (Storch and Thumser, 2000). LCFA uptake into adipocytes has also been shown to depend on fatty acid translocase (FAT/CD36) and the membrane fatty acid-binding protein (FABPpm) (Stremmel *et al.*, 2001). These proteins operate in the context of lipid rafts, the highly ordered lipid microdomains distinguished by specific interaction between sterols and sphingo-lipids (Pohl *et al.*, 2004; Simons and Ikonen, 1997). It is possible that some of the emerging transporter molecules will be highly effective as targets of therapy due to their functional capacity to enable uptake of the molecules binding to them.

White adipose tissue is particularly interesting organ, cells of which respond to extracellular signals transmitted from the central nervous system and other organs in order to maintain energy homeostasis (Rosen and MacDougald, 2006). Adipose cell differentiation is wired with the molecular machinery required for the uptake of nutrients, with glucose and fatty acids being the major resources for energy storage (Bederman *et al.*, 2009). The major type of fatty acids utilized by adipocytes is the LCFAs (McArthur *et al.*, 1999). Our unpublished studies have uncovered prohibitin (Phb) expression on adipocyte cell surface in addition to its previously reported marking of adipose vasculature (Kolonin *et al.*, 2004). Phb is a multifunctional protein originally identified as a proliferation inhibitor (McClung *et al.*, 1995). It has been later characterized as a component of the mitochondrial membrane, where it is involved in the assembly of multiprotein complexes (Nijtmans *et al.*, 2000). Phb, expressed in the form of several isoforms and posttranslational modifications, localizes to various com-partments within the cell, including the cell membrane (Mishra *et al.*, 2005; Rajalingam *et al.*, 2005), and is also produced by tumor cells and secreted (Mengwasser *et al.*, 2004; Wang *et al.*, 2004). Correspondingly, the array of the Phb's diverse functions has grown to span from regulating proapoptotic tran-scriptional machinery in the nucleus to serving as a cell surface receptor for infectious microorganisms (Fusaro *et al.*, 2003; Sharma and Qadri, 2004). Our unpublished data suggest that expression of Phb on adipocyte surface may contribute to the effectiveness of previously reported Phb-directed adipose tissue targeting.

Identification of Phb as a component of lipid rafts in association with other integral membrane receptor complexes and its isolation from macrophage phagosome proteome has pointed to a possible role of Phb in endocytosis (Garin et al., 2001). Caveolin is another lipid raft protein that has originally been characterized as an endothelial target but later also found overexpressed in other tissues including tumors (McIntosh et al., 2002). In white adipose tissue, Caveolin-1 is overexpressed in adipocyte cell membrane (Kampf et al., 2007) and, although its extracellular exposure is still debated, has been shown to move to cell membrane in adipocytes (Huo et al., 2003). The scavenger receptor CD36 is also expressed in adipocyte but not preadipocyte plasma membranes (Kampf et al., 2007). Finally, integrin $\alpha 6$ (Liu et al., 2005) appears to represent another adipocyte cell surface marker, as its complex with $\beta 1$ integrin replaces the $\alpha 5$ integrin partnering with $\beta 1$ integrin in adipose progenitor cells during adipogenesis (Liu et al., 2005).

In the heart, the endocardial endothelium has been shown to become permeable under pathological conditions. Therefore, cardiomyocyte-specific treatments and diagnostic probes have a potential to be effective (Brutsaert and Andries, 1992). Interestingly, selection of peptides homing to cardiomyocytes in vivo has demonstrated that the proteins expressed by these cells tend to also be expressed by the heart endothelium (McGuire et al., 2004; Zhang et al., 2005). A protein previously unappreciated as a prospective cardiomyocyte marker is the interleukin-1 receptor family member ST2. For many years, ST2 has been studied in the context of inflammatory and autoimmune disease as an orphan receptor. Identification of interleukin-33 (IL-33) as a functional ligand for ST2 has uncovered the role of IL-33/ST2 signaling not only in T-cell maturation but also in cardiovascular disease. The IL-33/ST2 interaction mediating the fibroblast–cardiomyocyte communication may represent a promising approach to deliver treatment to heart cells and establish a therapeutic target for the prevention of heart failure (Kakkar and Lee, 2008). Isolation of peptides homing to not only heart muscle but also skeletal muscle cells has been made possible by phage display library screening (Samoylova and Smith, 1999). Although the receptors for the ligand peptides have not been identified, this study proves the principle that not only cardiac but also skeletal myocytes are accessible to systemic circulation, which can be exploited for targeted delivery of therapy to muscle cells directly.

Placenta is the organ in which the blood directly contacts with its functional cells, thus being an outlier in terms of the endothelium posing a barrier for deep tissue access (Jollie, 1990). Maternofetal molecule exchange occurs by filtration of blood from the maternal to the fetal side of the placenta through several distinct cell layers of the villous placental epithelium composed of trophoblast-derived cells (Knobil and Neil, 1994). Although the structure of the placenta is different in primates and rodents, labyrinthine trophoblast of the mouse chorioallantoic placenta, containing syncytial cells, is analogous to the human villous syncytiotrophoblast, thus offering the mouse as a relevant model of human disease

(Burton and Watson, 1997). While it is clear that proper placental transport is critical for normal pregnancy progression, there is limited understanding of the precise molecular mechanisms regulating maternofetal transport (Goldenberg et al., 2001). Embryotoxins that may block receptors required for transportation of nutrients to the fetus apparently target the chorioallantoic placenta; however, the roles of specific amino acid, fatty acid, nucleoside, and carbohydrate transporters expressed in placenta have not been established (Chandorkar et al., 1999; Knipp et al., 1999; Maranghi et al., 1998). While existing animal models are certainly appropriate for testing pregnancy hazards (Elovitz and Mrinalini, 2004; Foidart et al., 1983; Hirsch and Wang, 2005; Knezevic et al., 1999), there have been surprisingly few studies attempting to identify individual components of the molecular machinery regulating placental transcytosis that may mediate the pregnancy complications.

We have previously used the in vivo phage display technology to isolate ligands that bind to receptors expressed on the villous placental epithelium. We established that one of these peptides, TPKTSVT, mimics a domain of FcRn and binds to β2-microglobulin, thus disrupting the biological FcRn/β2m interaction at the maternofetal interface (Kolonin et al., 2002). As a follow-up of this work, we have surveyed other placenta-homing peptide sequences for similarity to known proteins that could function in the context of placental epithelium. According to our unpublished results, some of these peptide motifs appear to mimic proteins interacting with receptors involved in amino acid transport, whereas other motifs may home to components of ECM-degraded MMPs or to coreceptors of MHC molecules similar to β2m (Table 3.2). Interestingly, it has been noted previously that such receptors capable of internalizing their peptide ligands often serve as docking receptors for viruses (Kontoyiannis et al., 2003). Consistent with this notion, we identified similarity of the same peptide motifs to coat proteins of mammalian viruses. Similarity of the placenta-homing motifs to segments of viral coat proteins may not only provide valuable information on the identity of targeted receptors but also establish molecular mechanisms mediating placental viral infections, possible role of which in pregnancy complications has not been explored.

E. Hematopoietic cell markers

The bone marrow, with its highly fenestrated microvasculature and slow circulation, provides a striking example of an organ in which various types of hematopoietic progenitor cells are directly exposed to blood. Other hematopoietic organs, such as spleen, thymus, and lymph nodes, are enriched for certain populations of lymphocytes and myeloid cells undergoing site-specific homing and differentiation. Obviously, cells of these differentially distributed subpopulations, which undergo maturation and antigen recognition programs in the

Table 3.2. Candidate Proteins Potentially Mimicked by Peptides Homing to Placenta

Placenta-homing peptide motif	Similar motif found by BLAST in proteins (GeneBank accession)	
	Mammalian	Viral
TP<u>K</u>T<u>S</u>VT	PPKTTVT MHC-I (<u>AAD43175</u>)	TPHTSVT Herpesvirus protein TRL12 (<u>CAA35298</u>)
<u>R</u>APG<u>G</u>R	APGGVR L-type amino acid transporter 3 (<u>XP_508432</u>)	PGGVR Herpesvirus protein UL61 (<u>CAA35376</u>)
RM<u>D</u>G<u>P</u>R	RMDSP Glutamate receptor (<u>XP_914719</u>)	RMDSPV HIV-1 protein GAG (<u>AAL29386</u>)
<u>Y</u>IRP<u>F</u>T<u>L</u>	LTFTRVY Matrix metalloproteinase 9 (<u>NP_038627</u>)	YVRPFT Adenovirus protein E3/49K (<u>AAM43819</u>)
<u>L</u>G<u>LRSVG</u>	LGIPNIG MHC Class II (<u>1718342A</u>)	GVSRLG Herpesvirus U21 protein EJLF2 (<u>Q69556</u>)

For sequence similarity search to mouse and viral proteins, peptide-containing motifs (in either orientation) were screened using the BLAST software (NCBI). Peptide motif amino acids identical to the corresponding mammalian or viral amino acids are underlined.

respective organs, are exposed to circulation. Capitalizing on the receptors exposed on individual lineages of blood cells, such as the known T-lymphocyte and B-lymphocyte markers, could be an effective approach to deliver agents to the respective lymphoid and myeloid organs in a directed way.

Various cells are mobilized from the bone marrow and possibly other organs in pathology and recruited by remote tissues (Kolonin and Simmons, 2009). Cells undergoing such pathological migration include various populations of leukocytes, as well as nonhematopoietic progenitors (Kucia *et al.*, 2004), such as endothelial progenitor cells (EPC) and possibly MSC. The capacity of various cell populations to "sense" sites of hypoxia/injury/inflammation, as well as tumors, has been demonstrated (Hall *et al.*, 2007). For example, macrophages are attracted by colony-stimulating factor-1 (CSF-1) produced by tumor cells and migrate to tumors where they produce various growth factors such as VEGF, which supports cancer vascularization. Recruitment of progenitor cells for vasculogenesis is a recently appreciated phenomenon taking place in cancer and other pathological conditions (Folkman, 2006). Several populations of circulating cells other than EPC have been recently implicated in supporting tumor vasculogenesis (Bertolini *et al.*, 2006; Goon *et al.*, 2006). Tumor-associated dendritic cells (TADCs), a new leukocyte population expressing both DC and endothelial markers, were shown to participate in the assembly of ovarian carcinoma neovasculature (Conejo-Garcia *et al.*, 2004). Recruited blood circulating cells (RBCCs), which express VEGFR1 but not VEGFR2, are driven by VEGF and SDF-1 to various organs and engage into neovasculogenesis (Grunewald *et al.*, 2006). Moreover, a population of TIE2-expressing monocytes (TEMs) was shown to be recruited to tumor sites and promote cancer angiogenesis in a paracrine manner after adhering to newly forming blood vessels (De Palma *et al.*, 2005). Some or all of these CD45[+] circulating proangiogenic cell populations may be identical to previously described fibrocytes (Hartlapp *et al.*, 2001). In benign tissues, injury and inflammatory responses are also well established as signals resulting in local concentration of histocytes, fibrocytes, and other types of cells engaging in repair. As far as lymphoid cells are concerned, adaptive immune response against tissue-specific antigens is one of the reasons why B cells may accumulate in a distinct site. For example, a unique subset of B cells, termed B2, was recently identified in the intestine (Shimomura *et al.*, 2008), and their markers may have clinical relevance.

Markers of differentially distributing blood cells, some of which are established, can serve as targets for selective delivery of agents to their destination organs. If a systemically administered agent happens to recognize a receptor selectively expressed on a particular type of blood cells concentrated in a certain tissue, this agent will also become concentrated in that tissue. Thus, even though not a part of the organ from the developmental point of view, circulating cells can be useful in imaging and therapeutic applications. For example, perivascular macrophages have been shown to upregulate the CD163 marker in certain pathological conditions (Borda *et al.*, 2008). In regard to organ-specific T- and B- lymphocyte subpopulations, individual members of the chemokine receptor family, differentially expressed as the mechanistic basis for cell homing, make a strong case. Identification of other markers for these cells, as well as for various types of bone marrow progenitors, such as stromal/pericyte progenitors, would enable efficient concentration of systemically administered agents in a specific organ.

F. Stem cell markers

A separate category of cells that is distinguished based on function rather than on microanatomical localization is stem/progenitor cells. A stem cell is unique in that it has the ability to self-renew indefinitely and to proliferate and differentiate into specialized cells under certain physiological or pathological conditions. Stem/progenitor cells have emerged as an attractive therapeutic target. Because the function of stem cells is to proliferate and to populate the organ with various cell types, the efficacy of therapy could be markedly enhanced by targeting stem cells, rather than differentiated cells, with gene correction vectors. Likewise, targeted delivery of cytotoxic agents to stem cells could be useful in certain clinical settings. Thus, directed delivery of gene therapy vectors to stem cells in disease in the future could become the solution to various genetic diseases.

Hematopoietic progenitors, primarily residing in the bone marrow, are certainly accessible by circulating molecules. Human HSC are identified in cells lacking hematopoietic lineage differentiation markers (Lin$^-$) and expressing CD34 and Thy-1 on the surface. Murine HSC have been identified as cKit$^+$, Sca1$^+$, Lin$^-$ cells, although other markers including Thy1.1, as well as flk2, CD105 (Endoglin), and CD150 (Slamf1), have refined their identity (Bryder et al., 2006). Using these markers to specifically target HSC, for example, for the purpose of skewing the hematopoietic program toward a certain lineage, would have a much more dramatic effect than similar treatment of downstream transit-amplifying cells. Although less abundant, endothelial progenitors represent a cell type with apparent importance in cancer progression, as they appear to serve as building blocks to the developing blood vessels at least in certain cancers. Because of their vascular integration, they may be particularly useful for certain gene therapy applications. For instance, secondary treatment with agents for induction of suicide genes delivered with EPC could be used for tumor vasculature ablation.

Mesenchymal stem cells have been explored to a comparatively less extent than the HSC and endothelial progenitors, both in terms of their distribution and in terms of markers present on them. It has been established that in vivo MSC can be distinguished from other cell types based on the expression of markers typical for mesenchymal cells (CD29, CD44, CD13, and CD90) as well as of CD34 (a marker long thought to be absent on mesenchymal cells due to its fast downregulation in culture) and combined with the lack of the hematopoietic marker CD45 and the endothelial marker CD31 (Gimble et al., 2007; Traktuev et al., 2008). Therefore, the CD34$^+$CD45$^-$CD31$^-$ immunophenotype is unique for primary MSC. The role of MSC in wound repair and tissue injury has been uncovered by a series of recent studies (Bianco et al., 2008; Stappenbeck and Miyoshi, 2009). While the role of MSC in organ regeneration is well acknowledged, the accumulating evidence indicates that MSC also represent a source of progenitors with pathological potential (Kolonin and Simmons, 2009). Due to these multiple roles of perivascular cells, they feature potentially important targets in various diseases. According to a number of studies by us and others, MSC, which are capable of homing to tumors, can promote cancer progression (Galie et al., 2007; Zhang et al., 2009). Based on these observations, MSC have been suggested to represent the progenitors of tumor stroma, commonly termed cancer-associated fibroblasts (Orimo and Weinberg, 2006; Wels et al., 2008). The secretion of angiogenic and antiapoptotic factors by MSC, as well as their ability to suppress T cell-mediated immune response (Chamberlain et al., 2007; Jones and McTaggart, 2008), shed light on the possible mechanisms underlying the stimulatory effect of MSC on cancer progression. MSC implication in cancer progression is one setting where the ability to inactivate this cell population would reduce their tropic effects on growing tumor. Another example of a pathological setting where MSC targeting could change the therapeutic outcome is obesity. Because they serve as cells giving rise to newly forming adipocytes, depleting MSC in white adipose tissue could become an approach to long-term obesity management. Despite the lack of specific markers, it is only a matter of time before MSC are fully appreciated as a therapeutic target.

A subject of intensive debate is the notion of "cancer stem cells" (CSC) and the possibility of their targeting. The cancer progenitor cell paradigm has emerged based on the notion that only a subset of tumor cells is capable of recreating the tumor upon transplantation, while maintaining the heterogeneity of cell types present within it (Clarke and Fuller, 2006; Reya et al., 2001). This observation is consistent with the existence of tumor-initiating cells that are commonly (although perhaps inappropriately) termed CSC (Rosen and Jordan, 2009). A number of molecules potentially useful as markers of CSC have been identified. In acute myeloid leukemia (AML), approximately one in a million blasts, corresponding to $CD34^+/CD38^-$ cell population, has been shown to possess tumor-initiating capacity (Dick and Lapidot, 2005). In multiple myeloma, the clonogenic potential has been demonstrated for cells negative for the cell surface marker syndecan-1 (CD138) in the $CD138^+$ bulk of tumor cells (Matsui et al., 2004). In solid tumors, evidence for the CSC existence has been first demonstrated for mammary carcinoma through the isolation of $CD44^+/CD24^-$ cells with clonogenic activity, which constituted 2% of the primary tumor cells (Ponti et al., 2006). The "stem cell marker" CD133 has been shown to identify the population of brain cancer cells capable of reconstituting tumors in mouse xenotransplantation models (Singh et al., 2004). CD133 was also used by Dick and colleagues to investigate human colon cancer cell population with tumor growth-initiating potency (O'Brien et al., 2007). Prostate cancer has also been extensively studied in attempts to demonstrate CSC-dependence of cancer progression (Nikitin et al., 2007; Tang et al., 2007). Witte and colleagues showed that urethra-proximal basal epithelium in mice contains a subpopulation of $Sca-1^+CD49f^+$ cells with a colony-forming potential 60-fold over other cell types (Lawson et al., 2007). These self-renewing prostate stem cells could be differentiated into typical prostate tubule structures containing both basal and luminal cells in vivo. It was recently shown that Sca-1 positive cells constitute a population of stem cell not only in normal mouse prostate but also in prostate cancer, and that p63 also plays a role in the prostate CSC function (Wang et al., 2006). As in the case of breast cancer, different groups have claimed different combinations of markers defining the clonogenic populations. A $CD44^+/CD133^+$ phenotype was reported for human prostate CSC, as well as for normal prostate stem cells (Collins et al., 2005; Richardson et al., 2004).

Interestingly, CD133 expression, detected in a subpopulation of prostate cancer cells, corresponded to the loss of androgen receptor in clinical prostate specimens (Miki et al., 2007). Also, a homologue of the Ly-6/Thy-1, named prostate stem cell antigen, was reported to be overexpressed in human transitional cell carcinoma (Amara et al., 2001). The marked discrepancies in the estimated CSC frequency in independent studies suggest that different populations of stem cells may be operating in different tumors, which is likely to feature a component of patient-specific variability.

Discovery of cell-surface markers for stem cells is an important component of current biomedicine pursued by numerous groups (Gundry et al., 2008). While a number of transcription factors, such as Oct4 and Nanog, are clearly marking stem cells, lack of cell surface exposure makes these molecules of limited practical use as therapeutic targets. Identification of cell surface markers of stem cells is in its dormancy. While molecules such as CD34 and CD133 have been used by many as pan-stem cell markers, their expression is by no means restricted to progenitor cells, with CD34 expressed in adult HSC and vasculature, whereas CD133 is in fact expressed in a wide range of differentiated epithelial cells in adult mouse tissues (Shmelkov et al., 2008). Notably, members of the CD34 family, namely CD34, podocalyxin, and endoglycan, show differential expression patterns, suggesting that they could be associated with tissue-specific progenitor populations (Nielsen and McNagny, 2008).

III. CONCLUDING REMARKS AND FUTURE DIRECTIONS

Individual tissue compartments may have different context-dependent modulating effects on the efficacy of therapeutic or diagnostic agent delivery. On one hand, when the receptor is expressed in more than one cell type in an organ, accumulation and retention of a receptor-binding agent may be more efficient and specific. This is clearly beneficial in the case of imaging agent targeting, as well is in therapeutic delivery when the bystander effect, favored by the receptor shared on adjacent cell types, helps the treatment to reach the target cells. In some cases, cytotoxic targeting of both endothelial and nonendothelial cells in a tissue could be highly effective as a multipronged approach to tissue ablation. In other cases, however, one could envision a scenario where, due to the targeting marker expression, a therapeutic vector is sequestered in a cell type that is not aimed to be affected. Such an unwanted decoy function of targeting receptors is even more so likely to diminish the payload effect when markers in the ECM are recognized by the targeted vector. This indeed may be a rather common situation, as numerous proteins expressed on vascular cells undergo processing and are secreted into the ECM either as full-length molecules or as alternative splicing derivatives or proteolytic fragments. Presence of such isoforms in the adjacent ECM is quite common, although systemic distribution or translocation to other tissues is also possible.

To a large extent, the progress in tissue targeting is impeded by the lack of markers. To this day, no specific cell surface markers have been identified for certain tissues/cell types. One example is the MSC, which currently cannot be distinguished from other populations of stromal cells or fibroblasts based on a single cell surface marker. In addition, even for markers currently used to prospectively identify or target certain cell populations, specificity is often only partial. Besides, there is an issue of species-dependent expression differences in markers, thus making it necessary to carefully evaluate the target molecules identified by using animal models in humans. Advances in genomics and proteomics may find to identify isoforms, such as those resulting from alternative splicing, that lead to different variants of a cell surface protein, which could be expressed in organ-specific patterns. In addition, because many types of protein modifications appear to take place in tissue-specific contexts, there are reasons to believe that today we are only looking at the tip of the iceberg of cell surface molecules. Glucosylation, lipidation, methylation, sulfation, phosphorylation, and various other types of modifications that the proteins are subjected to, in particularly upon secretion, create enormous diversity of isoforms and epimers. Some examples include the CD44 marker with its multiple subtypes, proteoglycans, such as decorin and CD13, which has been shown to serve as a tumor-specific marker (likely due to not yet established isoforms) despite of its ubiquitous expression (Curnis et al., 2002). Analyzing the expression selectivity for isoforms of most cell surface receptors is currently beyond our reach due to the lack of specific antibodies. In cases when identification of cell type-specific surface receptors remains challenging, it might be possible to capitalize on combinations of cell surface molecules. For instance, perivascular stromal cells can be identified as those expressing CD34 and mesenchymal markers such as CD44, CD13, CD90, and CD29. Decorating diagnostic or therapeutic vectors with homing moieties targeting both CD34 and a mesenchymal marker would render the vector a bimodal homing capacity and, therefore, selectivity for the target cell type. Such a strategy may provide a way to enrich the vector in the tissue of interest and should be considered in parallel with the search for new tissue-specific markers.

The clinical relevance of nonendothelial targets has only recently begun to surface. It is likely that capitalization on both endothelial markers and molecules expressed by other cell types will result in combinatorial approaches that will enhance the efficacy of organ-specific imaging or treatment.

Acknowledgment

I thank Alexis Daquinag for providing the images used in Fig. 3.3.

References

Amara, N., Palapattu, G. S., Schrage, M., Gu, Z., Thomas, G. V., Dorey, F., Said, J., and Reiter, R. E. (2001). Prostate stem cell antigen is overexpressed in human transitional cell carcinoma. *Cancer Res.* **61,** 4660–4665.

Arap, W., Pasqualini, R., and Ruoslahti, E. (1998). Cancer treatment by targeted drug delivery to tumor vasculature in a mouse model. *Science* **279,** 377–380.

Arap, W., Haedicke, W., Bernasconi, M., Kain, R., Rajotte, D., Krajewski, S., Ellerby, H. M., Bredesen, D. E., Pasqualini, R., and Ruoslahti, E. (2002a). Targeting the prostate for destruction through a vascular address. *Proc. Natl. Acad. Sci. USA* **99,** 1527–1531.

Arap, W., Kolonin, M. G., Trepel, M., Lahdenranta, J., Cardó-Vila, M., Giordano, R. J., Mintz, P. J., Ardelt, P. U., Yao, V. J., Vidal, C. I., et al. (2002b). Steps toward mapping the human vasculature by phage display. *Nat. Med.* **8,** 121–127.

Arap, M. A., Lahdenranta, J., Mintz, P. J., Hajitou, A., Sarkis, A. S., Arap, W., and Pasqualini, R. (2004). Cell surface expression of the stress response chaperone GRP78 enables tumor targeting by circulating ligands. *Cancer Cell* **6,** 275–284.

Baluk, P., Morikawa, S., Haskell, A., Mancuso, M., and McDonald, D. M. (2003). Abnormalities of basement membrane on blood vessels and endothelial sprouts in tumors. *Am. J. Pathol.* **163,** 1801–1815.

Barclay, A. N. (1981). Different reticular elements in rat lymphoid tissue identified by localization of Ia, Thy-1 and MRC OX 2 antigens. *Immunology* **44,** 727–736.

Bazan-Peregrino, M., Seymour, L. W., and Harris, A. L. (2007). Gene therapy targeting to tumor endothelium. *Cancer Gene Ther.* **14,** 117–127.

Bederman, I. R., Foy, S., Chandramouli, V., Alexander, J. C., and Previs, S. F. (2009). Triglyceride synthesis in epididymal adipose tissue: Contribution of glucose and non-glucose carbon sources. *J. Biol. Chem.* **284,** 6101–6108.

Bergers, G., Song, S., Meyer-Morse, N., Bergsland, E., and Hanahan, D. (2003). Benefits of targeting both pericytes and endothelial cells in the tumor vasculature with kinase inhibitors. *J. Clin. Invest.* **111,** 1287–1295.

Berk, P. D., and Stump, D. D. (1999). Mechanisms of cellular uptake of long chain free fatty acids. *Mol. Cell. Biochem.* **192,** 17–31.

Bertolini, F., Shaked, Y., Mancuso, P., and Kerbel, R. S. (2006). The multifaceted circulating endothelial cell in cancer: Towards marker and target identification. *Nat. Rev. Cancer* **6,** 835–845.

Bevilacqua, M. P., Stengelin, S., Gimbrone, M. A., Jr., and Seed, B. (1989). Endothelial leukocyte adhesion molecule 1: An inducible receptor for neutrophils related to complement regulatory proteins and lectins. *Science* **243,** 1160–1165.

Bianco, P., Robey, P. G., and Simmons, P. J. (2008). Mesenchymal stem cells: Revisiting history, concepts, and assays. *Cell Stem Cell* **2,** 313–319.

Biedermann, B. C. (2001). Vascular endothelium: Checkpoint for inflammation and immunity. *News Physiol. Sci.* **16,** 84–88.

Bissell, M. J., and Radisky, D. (2001). Putting tumours in context. *Nat. Rev. Cancer* **1,** 46–54.

Boda-Heggemann, J., Regnier-Vigouroux, A., and Franke, W. W. (2009). Beyond vessels: Occurrence and regional clustering of vascular endothelial (VE-)cadherin-containing junctions in non-endothelial cells. *Cell Tissue Res.* **335,** 49–65.

Boquest, A. C., Noer, A., and Collas, P. (2006). Epigenetic programming of mesenchymal stem cells from human adipose tissue. *Stem Cell Rev.* **2**, 319–329.

Borda, J. T., Alvarez, X., Mohan, M., Hasegawa, A., Bernardino, A., Jean, S., Aye, P., and Lackner, A. A. (2008). CD163, a marker of perivascular macrophages, is up-regulated by microglia in simian immunodeficiency virus encephalitis after haptoglobin-hemoglobin complex stimulation and is suggestive of breakdown of the blood-brain barrier. *Am. J. Pathol.* **172**, 725–737.

Bornstein, P., and Sage, E. H. (2002). Matricellular proteins: Extracellular modulators of cell function. *Curr. Opin. Cell Biol.* **14**, 608–616.

Bradshaw, A. D., Graves, D. C., Motamed, K., and Sage, E. H. (2003). SPARC-null mice exhibit increased adiposity without significant differences in overall body weight. *Proc. Natl. Acad. Sci. USA* **100**, 6045–6050.

Brekken, R. A., and Thorpe, P. E. (2001). VEGF-VEGF receptor complexes as markers of tumor vascular endothelium. *J. Control Release* **74**, 173–181.

Brutsaert, D. L., and Andries, L. J. (1992). The endocardial endothelium. *Am. J. Physiol. Endocrinol. Metab.* **263**, 985–1002.

Bryder, D., Rossi, D. J., and Weissman, I. L. (2006). Hematopoietic stem cells: The paradigmatic tissue-specific stem cell. *Am. J. Pathol.* **169**, 338–346.

Burastero, S. E., Rossi, G. A., and Crimi, E. (1998). Selective differences in the expression of the homing receptors of helper lymphocyte subsets. *Clin. Immunol. Immunopathol.* **89**, 110–116.

Burg, M. A., Pasqualini, R., Arap, W., Ruoslahti, E., and Stallcup, W. B. (1999). NG2 proteoglycan-binding peptides target tumor neovasculature. *Cancer Res.* **59**, 2869–2874.

Burton, G. J., and Watson, A. L. (1997). The structure of the human placenta: Implications for initiating and defending against virus infections. *Rev. Med. Virol.* **7**, 219–228.

Buse, J. B., Yasuda, K., Lay, T. P., Seo, T. S., Olson, A. L., Pessin, J. E., Karam, J. H., Seino, S., and Bell, G. I. (1992). Human GLUT4/muscle-fat glucose-transporter gene. Characterization and genetic variation. *Diabetes* **41**, 1436–1445.

Cardo-Vila, M., Zurita, A. J., Giordano, R. J., Sun, J., Rangel, R., Guzman-Rojas, L., Anobom, C. D., Valente, A. P., Almeida, F. C., Lahdenranta, J., *et al.* (2008). A ligand peptide motif selected from a cancer patient is a receptor-interacting site within human interleukin-11. *PLoS ONE* **3**, 3452–3457.

Chamberlain, G., Fox, J., Ashton, B., and Middleton, J. (2007). Concise review: Mesenchymal stem cells: Their phenotype, differentiation capacity, immunological features, and potential for homing. *Stem Cells* **25**, 2739–2749.

Chamberlain, G., Wright, K., Rot, A., Ashton, B., and Middleton, J. (2008). Murine mesenchymal stem cells exhibit a restricted repertoire of functional chemokine receptors: Comparison with human. *PLoS ONE* **3**, 2934–2939.

Chandorkar, G. A., Ampasavate, C., Stobaugh, J. F., and Audus, K. L. (1999). Peptide transport and metabolism across the placenta. *Adv. Drug Deliv. Rev.* **38**, 59–67.

Charron, M. J., and Kahn, B. B. (1990). Divergent molecular mechanisms for insulin-resistant glucose transport in muscle and adipose cells *in vivo*. *J. Biol. Chem.* **265**, 7994–8000.

Chatterjee, S. K., and Zetter, B. R. (2005). Cancer biomarkers: Knowing the present and predicting the future. *Future Oncol.* **1**, 37–50.

Chin, Y. H., Cai, J. P., and Xu, X. M. (1991). Tissue-specific homing receptor mediates lymphocyte adhesion to cytokine-stimulated lymph node high endothelial venule cells. *Immunology* **74**, 478–483.

Choong, P. F. (2003). The molecular basis of skeletal metastases. *Clin. Orthop.* S19–S31.

Christian, S., Winkler, R., Helfrich, I., Boos, A. M., Besemfelder, E., Schadendorf, D., and Augustin, H. G. (2008). Endosialin (Tem1) is a marker of tumor-associated myofibroblasts and tumor vessel-associated mural cells. *Am. J. Pathol.* **172**, 486–494.

Chute, J. P. (2006). Stem cell homing. *Curr. Opin. Hematol.* **13**, 399–406.

Cinti, S. (2000). Anatomy of the adipose organ. *Eat. Weight Disord.* **5,** 132–142.

Clark, C. J., and Sage, E. H. (2008). A prototypic matricellular protein in the tumor microenvironment—Where there's SPARC, there's fire. *J. Cell. Biochem.* **104,** 721–732.

Clarke, M. F., and Fuller, M. (2006). Stem cells and cancer: Two faces of eve. *Cell* **124,** 1111–1115.

Collins, A. T., Berry, P. A., Hyde, C., Stower, M. J., and Maitland, N. J. (2005). Prospective identification of tumorigenic prostate cancer stem cells. *Cancer Res.* **65,** 10946–10951.

Conejo-Garcia, J. R., Benencia, F., Courreges, M. C., Kang, E., Mohamed-Hadley, A., Buckanovich, R. J., Holtz, D. O., Jenkins, A., Na, H., Zhang, L., *et al.* (2004). Tumor-infiltrating dendritic cell precursors recruited by a beta-defensin contribute to vasculogenesis under the influence of Vegf-A. *Nat. Med.* **10,** 950–958.

Crandall, D. L., Hausman, G. J., and Kral, J. G. (1997). A review of the microcirculation of adipose tissue: Anatomic, metabolic, and angiogenic perspectives. *Microcirculation* **4,** 211–232.

Crisan, M., Yap, S., Casteilla, L., Chen, C. W., Corselli, M., Park, T. S., Andriolo, G., Sun, B., Zheng, B., Zhang, L., *et al.* (2008). A perivascular origin for mesenchymal stem cells in multiple human organs. *Cell Stem Cell* **3,** 301–313.

Curnis, F., Sacchi, A., Borgna, L., Magni, F., Gasparri, A., and Corti, A. (2000). Enhancement of tumor necrosis factor alpha antitumor immunotherapeutic properties by targeted delivery to aminopeptidase N (CD13). *Nat. Biotechnol.* **18,** 1185–1190.

Curnis, F., Arrigoni, G., Sacchi, A., Fischetti, L., Arap, W., Pasqualini, R., and Corti, A. (2002). Differential binding of drugs containing the NGR Motif to CD13 isoforms in tumor vessels, epithelia, and myeloid cells. *Cancer Res.* **62,** 867–874.

De Palma, M., Venneri, M. A., Galli, R., Sergi Sergi, L., Politi, L. S., Sampaolesi, M., and Naldini, L. (2005). Tie2 identifies a hematopoietic lineage of proangiogenic monocytes required for tumor vessel formation and a mesenchymal population of pericyte progenitors. *Cancer Cell* **8,** 211–226.

De Ugarte, D. A., Alfonso, Z., Zuk, P. A., Elbarbary, A., Zhu, M., Ashjian, P., Benhaim, P., Hedrick, M. H., and Fraser, J. K. (2003). Differential expression of stem cell mobilization-associated molecules on multi-lineage cells from adipose tissue and bone marrow. *Immunol. Lett.* **89,** 267–270.

Dick, J. E., and Lapidot, T. (2005). Biology of normal and acute myeloid leukemia stem cells. *Int. J. Hematol.* **82,** 389–396.

Durr, E., Yu, J., Krasinska, K. M., Carver, L. A., Yates, J. R., Testa, J. E., Oh, P., and Schnitzer, J. E. (2004). Direct proteomic mapping of the lung microvascular endothelial cell surface *in vivo* and in cell culture. *Nat. Biotechnol.* **22,** 985–992.

Dvorak, H. F. (1986). Tumors: Wounds that do not heal. Similarities between tumor stroma generation and wound healing. *N. Engl. J. Med.* **315,** 1650–1659.

Edlund, M., Sung, S. Y., and Chung, L. W. (2004). Modulation of prostate cancer growth in bone microenvironments. *J. Cell. Biochem.* **91,** 686–705.

Ellerby, H. M., Arap, W., Ellerby, L. M., Kain, R., Andrusiak, R., Rio, G. D., Krajewski, S., Lombardo, C. R., Rao, R., Ruoslahti, E., *et al.* (1999). Anti-cancer activity of targeted proapoptotic peptides. *Nat. Med.* **5,** 1032–1038.

Ellerby, H. M., Lee, S., Ellerby, L. M., Chen, S., Kiyota, T., del Rio, G., Sugihara, G., Sun, Y., Bredesen, D. E., Arap, W., and Pasqualini, R. (2003). An artificially designed pore-forming protein with anti-tumor effects. *J. Biol. Chem.* **278,** 35311–35316.

Elovitz, M. A., and Mrinalini, C. (2004). Animal models of preterm birth. *Trends Endocrinol. Metab.* **15,** 479–487.

Engwegen, J. Y., Gast, M. C., Schellens, J. H., and Beijnen, J. H. (2006). Clinical proteomics: Searching for better tumour markers with SELDI-TOF mass spectrometry. *Trends Pharmacol. Sci.* **27,** 251–259.

Essler, M., and Ruoslahti, E. (2002). Molecular specialization of breast vasculature: A breast-homing phage-displayed peptide binds to aminopeptidase P in breast vasculature. *Proc. Natl. Acad. Sci. USA* **99,** 2252–2257.

Fabbri, M., Bianchi, E., Fumagalli, L., and Pardi, R. (1999). Regulation of lymphocyte traffic by adhesion molecules. *Inflamm. Res.* **48,** 239–246.

Ferrara, N., and Kerbel, R. S. (2005). Angiogenesis as a therapeutic target. *Nature* **438,** 967–974.

Fidler, I. J. (1997). Molecular biology of cancer: Invasion and metastasis. *In* "Cancer: Principles and Practice of Oncology" (V. T. DeVita, S. Hellman, and S. A. Rosenberg, eds.), pp. 135–152. Lippincott-Raven, Philadelphia, PA.

Foidart, J. M., Yaar, M., Figueroa, A., Wilk, A., Brown, K. S., and Liotta, L. A. (1983). Abortion in mice induced by intravenous injections of antibodies to type IV collagen or laminin. *Am. J. Pathol.* **110,** 346–357.

Folkman, J. (2006). Angiogenesis. *Annu. Rev. Med.* **57,** 1–18.

Folkman, J. (2007). Angiogenesis: An organizing principle for drug discovery? *Nat. Rev. Drug Discov.* **6,** 273–286.

Framson, P. E., and Sage, E. H. (2004). SPARC and tumor growth: Where the seed meets the soil? *J. Cell. Biochem.* **92,** 679–690.

Friedenstein, A. J. (1980). Stromal mechanisms of bone marrow: Cloning *in vitro* and retransplantation *in vivo*. *Haematol. Blood Transfus.* **25,** 19–29.

Fusaro, G., Dasgupta, P., Rastogi, S., Joshi, B., and Chellappan, S. (2003). Prohibitin induces the transcriptional activity of p53 and is exported from the nucleus upon apoptotic signaling. *J. Biol. Chem.* **278,** 47853–47861.

Galie, M., Konstantinidou, G., Peroni, D., Scambi, I., Marchini, C., Lisi, V., Krampera, M., Magnani, P., Merigo, F., Montani, M., *et al.* (2007). Mesenchymal stem cells share molecular signature with mesenchymal tumor cells and favor early tumor growth in syngeneic mice. *Oncogene* **27,** 2542–2545.

Garin, J., Diez, R., Kieffer, S., Dermine, J. F., Duclos, S., Gagnon, E., Sadoul, R., Rondeau, C., and Desjardins, M. (2001). The phagosome proteome: Insight into phagosome functions. *J. Cell. Biol.* **152,** 165–180.

Gimble, J. M., Katz, A. J., and Bunnell, B. A. (2007). Adipose-derived stem cells for regenerative medicine. *Circ. Res.* **100,** 1249–1260.

Giordano, R. J., Cardó-Vila, M., Lahdenranta, J., Pasqualini, R., and Arap, W. (2001). Biopanning and rapid analysis of selective interactive ligands. *Nat. Med.* **7,** 1249–1253.

Goldenberg, R. L., Iams, J. D., Mercer, B. M., Meis, P. J., Moawad, A., Das, A., Miodovnik, M., Vandorsten, P. J., Caritis, S. N., Thurnau, G., and Dombrowski, M. P. (2001). The preterm prediction study: Toward a multiple-marker test for spontaneous preterm birth. *Am. J. Obstet. Gynecol.* **185,** 643–651.

Goon, P. K., Lip, G. Y., Boos, C. J., Stonelake, P. S., and Blann, A. D. (2006). Circulating endothelial cells, endothelial progenitor cells, and endothelial microparticles in cancer. *Neoplasia* **8,** 79–88.

Greene, A. K., Wiener, S., Puder, M., Yoshida, A., Shi, B., Perez-Atayde, A. R., Efstathiou, J. A., Holmgren, L., Adamis, A. P., Rupnick, M., *et al.* (2003). Endothelial-directed hepatic regeneration after partial hepatectomy. *Ann. Surg.* **237,** 530–535.

Griffin, N. M., and Schnitzer, J. E. (2008). Proteomic mapping of the vascular endothelium *in vivo* for vascular targeting. *Methods Enzymol.* **445,** 177–208.

Grunewald, M., Avraham, I., Dor, Y., Bachar-Lustig, E., Itin, A., Jung, S., Chimenti, S., Landsman, L., Abramovitch, R., and Keshet, E. (2006). VEGF-induced adult neovascularization: Recruitment, retention, and role of accessory cells. *Cell* **124,** 175–189.

Gunaratnam, M., Swank, S., Haider, S. M., Galesa, K., Reszka, A. P., Beltran, M., Cuenca, F., Fletcher, J. A., and Neidle, S. (2009). Targeting human gastrointestinal stromal tumor cells with a quadruplex-binding small molecule. *J. Med. Chem.* **52,** 3774–3783.

Gundry, R. L., Boheler, K. R., Van Eyk, J. E., and Wollscheid, B. (2008). A novel role for proteomics in the discovery of cell-surface markers on stem cells: Scratching the surface. *Proteomics Clin. Appl.* **2,** 892–903.

Hafner, C., Schmitz, G., Meyer, S., Bataille, F., Hau, P., Langmann, T., Dietmaier, W., Landthaler, M., and Vogt, T. (2004). Differential gene expression of Eph receptors and ephrins in benign human tissues and cancers. *Clin. Chem.* **50**, 490–499.

Hajitou, A., Trepel, M., Lilley, C. E., Soghomonyan, S., Alauddin, M. M., Marini, F. C., 3rd, Restel, B. H., Ozawa, M. G., Moya, C. A., Rangel, R., *et al.* (2006). A hybrid vector for ligand-directed tumor targeting and molecular imaging. *Cell* **125**, 385–398.

Hall, B., Dembinski, J., Sasser, A. K., Studeny, M., Andreeff, M., and Marini, F. (2007). Mesenchymal stem cells in cancer: Tumor-associated fibroblasts and cell-based delivery vehicles. *Int. J. Hematol.* **86**, 8–16.

Hallmann, R., Horn, N., Selg, M., Wendler, O., Pausch, F., and Sorokin, L. M. (2005). Expression and function of laminins in the embryonic and mature vasculature. *Physiol. Rev.* **85**, 979–1000.

Hanahan, D., and Weinberg, R. A. (2000). The hallmarks of cancer. *Cell* **100**, 57–70.

Hartlapp, I., Abe, R., Saeed, R. W., Peng, T., Voelter, W., Bucala, R., and Metz, C. N. (2001). Fibrocytes induce an angiogenic phenotype in cultured endothelial cells and promote angiogenesis *in vivo*. *FASEB J.* **15**, 2215–2224.

Hausman, G. J., and Richardson, R. L. (2004). Adipose tissue angiogenesis. *J. Anim. Sci.* **82**, 925–934.

Hausman, D. B., DiGirolamo, M., Bartness, T. J., Hausman, G. J., and Martin, R. J. (2001). The biology of white adipocyte proliferation. *Obes. Rev.* **2**, 239–254.

Hermani, A., Hess, J., De Servi, B., Medunjanin, S., Grobholz, R., Trojan, L., Angel, P., and Mayer, D. (2005). Calcium-binding proteins S100A8 and S100A9 as novel diagnostic markers in human prostate cancer. *Clin. Cancer Res.* **11**, 5146–5152.

Hirsch, E., and Wang, H. (2005). The molecular pathophysiology of bacterially induced preterm labor: Insights from the murine model. *J. Soc. Gynecol. Investig.* **12**, 145–155.

Hoffman, J. A., Giraudo, E., Singh, M., Zhang, L., Inoue, M., Porkka, K., Hanahan, D., and Ruoslahti, E. (2003). Progressive vascular changes in a transgenic mouse model of squamous cell carcinoma. *Cancer Cell* **4**, 383–391.

Hong, F. D., and Clayman, G. L. (2000). Isolation of a peptide for targeted drug delivery into human head and neck solid tumors. *Cancer Res.* **60**, 6551–6556.

Honn, K. V., and Tang, D. G. (1992). Adhesion molecules and tumor cell interaction with endothelium and subendothelial matrix. *Cancer Metastasis Rev.* **11**, 353–375.

Huo, H., Guo, X., Hong, S., Jiang, M., Liu, X., and Liao, K. (2003). Lipid rafts/caveolae are essential for insulin-like growth factor-1 receptor signaling during 3T3–L1 preadipocyte differentiation induction. *J. Biol. Chem.* **278**, 11561–11569.

Hynes, R. O. (2002). Integrins: Bidirectional, allosteric signaling machines. *Cell* **110**, 673–687.

Ikegawa, S. (2008). Expression, regulation and function of asporin, a susceptibility gene in common bone and joint diseases. *Curr. Med. Chem.* **15**, 724–728.

Jollie, W. P. (1990). Development, morphology, and function of the yolk-sac placenta of laboratory rodents. *Teratology* **41**, 361–381.

Jones, B. J., and McTaggart, S. J. (2008). Immunosuppression by mesenchymal stromal cells: From culture to clinic. *Exp. Hematol.* **36**, 733–741.

Joyce, J. A., Laakkonen, P., Bernasconi, M., Bergers, G., Ruoslahti, E., and Hanahan, D. (2003). Stage-specific vascular markers revealed by phage display in a mouse model of pancreatic islet tumorigenesis. *Cancer Cell* **4**, 393–403.

Kakkar, R., and Lee, R. T. (2008). The IL-33/ST2 pathway: Therapeutic target and novel biomarker. *Nat. Rev. Drug Discov.* **7**, 827–840.

Kampf, J. P., Parmley, D., and Kleinfeld, A. M. (2007). Free fatty acid transport across adipocytes is mediated by an unknown membrane protein pump. *Am. J. Physiol. Endocrinol. Metab.* **293**, 1207–1214.

Katz, E. B., Stenbit, A. E., Hatton, K., DePinho, R., and Charron, M. J. (1995). Cardiac and adipose tissue abnormalities but not diabetes in mice deficient in GLUT4. *Nature* **377**, 151–155.

Keller, E. T., Zhang, J., Cooper, C. R., Smith, P. C., McCauley, L. K., Pienta, K. J., and Taichman, R. S. (2001). Prostate carcinoma skeletal metastases: Cross-talk between tumor and bone. *Cancer Metastasis Rev.* **20**, 333–349.

Knezevic, N., Nikolic, B., Brajsa, K., Spaventi, R., Jonjic, N., Jonjic, S., and Marusic, S. (1999). Monoclonal antibodies against maternal major histocompatibility complex class I molecules induce rapid abortion in mice. *Am. J. Reprod. Immunol.* **41**, 217–223.

Knipp, G. T., Audus, K. L., and Soares, M. J. (1999). Nutrient transport across the placenta. *Adv. Drug Deliv. Rev.* **38**, 41–58.

Knobil, E., and Neil, J. D. (1994). Physiology of Reproduction, 2nd edn. Raven Press, New York.

Koivunen, E., Arap, W., Valtanen, H., Rainisalo, A., Medina, O. P., Heikkila, P., Kantor, C., Gahmberg, C. G., Salo, T., Konttinen, Y. T., *et al.* (1999). Tumor targeting with a selective gelatinase inhibitor. *Nat. Biotechnol.* **17**, 768–774.

Kolonin, M. G., and Simmons, P. J. (2009). Combinatorial stem cell mobilization. *Nat. Biotechnol.* **27**, 252–253.

Kolonin, M. G., Pasqualini, R., and Arap, W. (2001). Molecular addresses in blood vessels as targets for therapy. *Curr. Opin. Chem. Biol.* **5**, 308–313.

Kolonin, M. G., Pasqualini, R., and Arap, W. (2002). Teratogenicity induced by targeting a placental immunoglobulin transporter. *Proc. Natl. Acad. Sci. USA* **99**, 13055–13060.

Kolonin, M. G., Pasqualini, R., and Arap, W. (2003). Mapping human vascular heterogeneity by *in vivo* phage display. *In* "Genetics of Angiogenesis" (J. B. Hoying, ed.), BIOS Scientific Publishers Ltd., Oxford.

Kolonin, M. G., Saha, P. K., Chan, L., Pasqualini, R., and Arap, W. (2004). Reversal of obesity by targeted ablation of adipose tissue. *Nat. Med.* **10**, 625–632.

Kolonin, M. G., Bover, L., Sun, J., Zurita, A. J., Do, K. A., Lahdenranta, J., Cardó-Vila, M., Giordano, R. J., Jaalouk, D. E., Ozawa, M. G., *et al.* (2006a). Ligand-directed surface profiling of human cancer cells with combinatorial peptide libraries. *Cancer Res.* **66**, 34–40.

Kolonin, M. G., Sun, J., Do, K. A., Vidal, C. I., Ji, Y., Baggerly, K. A., Pasqualini, R., and Arap, W. (2006b). Synchronous selection of homing peptides for multiple tissues by *in vivo* phage display. *FASEB J.* **20**, 979–981.

Kontoyiannis, D. P., Pasqualini, R., and Arap, W. (2003). Aminopeptidase N inhibitors and SARS. *Lancet* **361**, 1558.

Koukourakis, M. I., Giatromanolaki, A., Harris, A. L., and Sivridis, E. (2006). Comparison of metabolic pathways between cancer cells and stromal cells in colorectal carcinomas: A metabolic survival role for tumor-associated stroma. *Cancer Res.* **66**, 632–637.

Krag, D. N., Shukla, G. S., Shen, G. P., Pero, S., Ashikaga, T., Fuller, S., Weaver, D. L., Burdette-Radoux, S., and Thomas, C. (2006). Selection of tumor-binding ligands in cancer patients with phage display libraries. *Cancer Res.* **66**, 7724–7733.

Kubes, P., and Kerfoot, S. M. (2001). Leukocyte recruitment in the microcirculation: The rolling paradigm revisited. *News Physiol. Sci.* **16**, 76–80.

Kucia, M., Ratajczak, J., Reca, R., Janowska-Wieczorek, A., and Ratajczak, M. Z. (2004). Tissue-specific muscle, neural and liver stem/progenitor cells reside in the bone marrow, respond to an SDF-1 gradient and are mobilized into peripheral blood during stress and tissue injury. *Blood Cells Mol. Dis.* **32**, 52–57.

Kucia, M., Reca, R., Miekus, K., Wanzeck, J., Wojakowski, W., Janowska-Wieczorek, A., Ratajczak, J., and Ratajczak, M. Z. (2005). Trafficking of normal stem cells and metastasis of cancer stem cells involve similar mechanisms: Pivotal role of the SDF-1-CXCR4 axis. *Stem Cells* **23**, 879–894.

Lawson, D. A., Xin, L., Lukacs, R. U., Cheng, D., and Witte, O. N. (2007). Isolation and functional characterization of murine prostate stem cells. *Proc. Natl. Acad. Sci. USA* **104**, 181–186.

Lebeau, A. M., Brennen, W. N., Aggarwal, S., and Denmeade, S. R. (2009). Targeting the cancer stroma with a fibroblast activation protein-activated promelittin protoxin. *Mol. Cancer Ther.* **8**(5), 1378–1386.

Lee, R. H., Kim, B., Choi, I., Kim, H., Choi, H. S., Suh, K., Bae, Y. C., and Jung, J. S. (2004). Characterization and expression analysis of mesenchymal stem cells from human bone marrow and adipose tissue. *Cell Physiol. Biochem.* **14**, 311–324.

Lee, J. H., Chung, K. Y., Bang, D., and Lee, K. H. (2006). Searching for aging-related proteins in human dermal microvascular endothelial cells treated with anti-aging agents. *Proteomics* **6**, 1351–1361.

Liotta, L. A., Ferrari, M., and Petricoin, E. (2003). Clinical proteomics: Written in blood. *Nature* **425**, 905.

Liu, Y., and Deisseroth, A. (2006). Tumor vascular targeting therapy with viral vectors. *Blood* **107**, 3027–3033.

Liu, J., DeYoung, S. M., Zhang, M., Zhang, M., Cheng, A., and Saltiel, A. R. (2005). Changes in integrin expression during adipocyte differentiation. *Cell Metab.* **2**, 165–177.

Lu, C., Kamat, A. A., Lin, Y. G., Merritt, W. M., Landen, C. N., Kim, T. J., Spannuth, W., Arumugam, T., Han, L. Y., Jennings, N. B., *et al.* (2007). Dual targeting of endothelial cells and pericytes in antivascular therapy for ovarian carcinoma. *Clin. Cancer Res.* **13**, 4209–4217.

Mackenzie, I. C., and Dabelsteen, E. (1987). Connective tissue influences on the expression of epithelial cell-surface antigens. *Cell Tissue Res.* **248**, 137–141.

Maihle, N. J., Baron, A. T., Barrette, B. A., Boardman, C. H., Christensen, T. A., Cora, E. M., Faupel-Badger, J. M., Greenwood, T., Juneja, S. C., Lafky, J. M., *et al.* (2002). EGF/ErbB receptor family in ovarian cancer. *Cancer Treat. Res.* **107**, 247–258.

Maranghi, F., Macri, C., Ricciardi, C., Stazi, A. V., and Mantovani, A. (1998). Evaluation of the placenta: Suggestions for a greater role in developmental toxicology. *Adv. Exp. Med. Biol.* **444**, 129–136.

Marchió, S., Lahdenranta, J., Schlingemann, R. O., Valdembri, D., Wesseling, P., Arap, M. A., Hajitou, A., Ozawa, M. G., Trepel, M., Giordano, R. J., *et al.* (2004). Aminopeptidase A is a functional target in angiogenic blood vessels. *Cancer Cell* **5**, 151–162.

Matsui, W., Huff, C. A., Wang, Q., Malehorn, M. T., Barber, J., Tanhehco, Y., Smith, B. D., Civin, C. I., and Jones, R. J. (2004). Characterization of clonogenic multiple myeloma cells. *Blood* **103**, 2332–2336.

Matsuyama, S., Ichihara-Tanaka, K., and Sekiguchi, K. (1994). Targeting of the immunoglobulin-binding domain of protein A to the extracellular matrix using a minifibronectin expression vector. *J. Biochem.* **116**, 898–904.

McArthur, M. J., Atshaves, B. P., Frolov, A., Foxworth, W. D., Kier, A. B., and Schroeder, F. (1999). Cellular uptake and intracellular trafficking of long chain fatty acids. *J. Lipid Res.* **40**, 1371–1383.

McCluggage, W. G., Sumathi, V. P., and Maxwell, P. (2001). CD10 is a sensitive and diagnostically useful immunohistochemical marker of normal endometrial stroma and of endometrial stromal neoplasms. *Histopathology* **39**, 273–278.

McClung, J. K., Jupe, E. R., Liu, X. T., and Dell'Orco, R. T. (1995). Prohibitin: Potential role in senescence, development, and tumor suppression. *Exp. Gerontol.* **30**, 99–124.

McDonald, D. M., and Foss, A. J. (2000). Endothelial cells of tumor vessels: Abnormal but not absent. *Cancer Metastasis Rev.* **19**, 109–120.

McGuire, M. J., Samli, K. N., Johnston, S. A., and Brown, K. C. (2004). In vitro selection of a peptide with high selectivity for cardiomyocytes in vivo. *J. Mol. Biol.* **342**, 171–182.

McIntosh, D. P., Tan, X. Y., Oh, P., and Schnitzer, J. E. (2002). Targeting endothelium and its dynamic caveolae for tissue-specific transcytosis in vivo: A pathway to overcome cell barriers to drug and gene delivery. *Proc. Natl. Acad. Sci. USA* **99**, 1996–2001.

Mengwasser, J., Piau, A., Schlag, P., and Sleeman, J. P. (2004). Differential immunization identifies PHB1/PHB2 as blood-borne tumor antigens. *Oncogene* **23**, 7430–7435.

Middleton, J., Patterson, A. M., Gardner, L., Schmutz, C., and Ashton, B. A. (2002). Leukocyte extravasation: Chemokine transport and presentation by the endothelium. *Blood* **100**, 3853–3860.

Miki, J., Furusato, B., Li, H., Gu, Y., Takahashi, H., Egawa, S., Sesterhenn, I. A., McLeod, D. G., Srivastava, S., and Rhim, J. S. (2007). Identification of putative stem cell markers, CD133 and CXCR4, in hTERT-immortalized primary nonmalignant and malignant tumor-derived human prostate epithelial cell lines and in prostate cancer specimens. *Cancer Res.* **67**, 3153–3561.

Mintz, P. J., Cardo-Vila, M., Ozawa, M. G., Hajitou, A., Rangel, R., Guzman-Rojas, L., Christianson, D. R., Arap, M. A., Giordano, R. J., Souza, G. R., et al. (2009). An unrecognized extracellular function for an intracellular adapter protein released from the cytoplasm into the tumor microenvironment. *Proc. Natl. Acad. Sci. USA* **106**, 2182–2187.

Mishra, S., Murphy, L. C., Nyomba, B. L., and Murphy, L. J. (2005). Prohibitin: A potential target for new therapeutics. *Trends Mol. Med.* **11**, 192–197.

Moghimi, S. M., Hunter, A. C., and Murray, J. C. (2005). Nanomedicine: Current status and future prospects. *FASEB J.* **19**, 311–330.

Mohle, R., Bautz, F., Rafii, S., Moore, M. A., Brugger, W., and Kanz, L. (1999). Regulation of transendothelial migration of hematopoietic progenitor cells. *Ann. NY Acad. Sci.* **872**, 176–185.

Molldrem, J. J., Komanduri, K., and Wieder, E. (2002). Overexpressed differentiation antigens as targets of graft-versus-leukemia reactions. *Curr. Opin. Hematol.* **9**, 503–508.

Morikawa, S., Baluk, P., Kaidoh, T., Haskell, A., Jain, R. K., and McDonald, D. M. (2002). Abnormalities in pericytes on blood vessels and endothelial sprouts in tumors. *Am. J. Pathol.* **160**, 985–1000.

Nie, J., Chang, B., Traktuev, D. O., Sun, J., March, K. L., Chan, L., Sage, E. H., Pasqualini, R., Arap, W., and Kolonin, M. G. (2008). Combinatorial peptides identify $\alpha5\beta1$ integrin as a receptor for the matricellular protein SPARC on adipose stromal cells. *Stem Cells* **26**, 2735–2745.

Nielsen, J. S., and McNagny, K. M. (2008). Novel functions of the CD34 family. *J. Cell Sci.* **121**, 3683–3692.

Nijtmans, L. G., de Jong, L., Artal Sanz, M., Coates, P. J., Berden, J. A., Back, J. W., Muijsers, A. O., van der Spek, H., and Grivell, L. A. (2000). Prohibitins act as a membrane-bound chaperone for the stabilization of mitochondrial proteins. *EMBO J.* **19**, 2444–2451.

Nikitin, A. Y., Matoso, A., and Roy-Burman, P. (2007). Prostate stem cells and cancer. *Histol. Histopathol.* **22**, 1043–1049.

Noel, D., Caton, D., Roche, S., Bony, C., Lehmann, S., Casteilla, L., Jorgensen, C., and Cousin, B. (2008). Cell specific differences between human adipose-derived and mesenchymal-stromal cells despite similar differentiation potentials. *Exp. Cell Res.* **314**, 1575–1584.

O'Brien, C. A., Pollett, A., Gallinger, S., and Dick, J. E. (2007). A human colon cancer cell capable of initiating tumour growth in immunodeficient mice. *Nature* **445**, 106–110.

Orimo, A., and Weinberg, R. A. (2006). Stromal fibroblasts in cancer: A novel tumor-promoting cell type. *Cell Cycle* **5**, 1597–1601.

Ozerdem, U. (2006). Targeting of pericytes diminishes neovascularization and lymphangiogenesis in prostate cancer. *Prostate* **66**, 294–304.

Paez-Ribes, M., Allen, E., Hudock, J., Takeda, T., Okuyama, H., Vinals, F., Inoue, M., Bergers, G., Hanahan, D., and Casanovas, O. (2009). Antiangiogenic therapy elicits malignant progression of tumors to increased local invasion and distant metastasis. *Cancer Cell* **15**, 220–231.

Pasqualini, R., and Arap, W. (2002). Vascular targeting. *In* "The Encyclopedia of Cancer" (J. R. Bertino, ed.), pp. 525–530. Academic Press, New Brunswick, NJ.

Pasqualini, R., and Ruoslahti, E. (1996). Organ targeting *in vivo* using phage display peptide libraries. *Nature* **380**, 364–366.

Pasqualini, R., Bourdoulous, S., Koivunen, E., Woods, V. L., Jr., and Ruoslahti, E. (1996). A polymeric form of fibronectin has antimetastatic effects against multiple tumor types. *Nat. Med.* **2**, 1197–1203.

Pasqualini, R., Koivunen, E., Kain, R., Lahdenranta, J., Sakamoto, M., Stryhn, A., Ashmun, R. A., Shapiro, L. H., Arap, W., and Ruoslahti, E. (2000). Aminopeptidase N is a receptor for tumor-homing peptides and a target for inhibiting angiogenesis. *Cancer Res.* **60,** 722–727.

Pasqualini, R., Arap, W., Rajotte, D., and Ruoslahti, E. (2001). In vivo selection of phage-display libraries. In "Phage Display: A Laboratory Manual" (C. Barbas, D. Burton, G. Silverman, and J. Scott, eds.), pp. 22.21–22.24. Cold Spring Harbor Laboratory Press, New York, NY.

Pietras, K., and Hanahan, D. (2005). A multitargeted, metronomic, and maximum-tolerated dose "chemo-switch" regimen is antiangiogenic, producing objective responses and survival benefit in a mouse model of cancer. *J. Clin. Oncol.* **23,** 939–952.

Pittenger, M. F., Mackay, A. M., Beck, S. C., Jaiswal, R. K., Douglas, R., Mosca, J. D., Moorman, M. A., Simonetti, D. W., Craig, S., and Marshak, D. R. (1999). Multilineage potential of adult human mesenchymal stem cells. *Science* **284,** 143–147.

Plopper, G. (2007). The extracellular matrix and cell adhesion. In "Cells" (B. Lewin, L. Cassimeris, V. Lingappa, and G. Plopper, eds.), pp. 645–702. Jones & Bartlett, Sudbury, MA.

Pohl, J., Ring, A., Ehehalt, R., Schulze-Bergkamen, H., Schad, A., Verkade, P., and Stremmel, W. (2004). Long-chain fatty acid uptake into adipocytes depends on lipid raft function. *Biochemistry* **43,** 4179–4187.

Ponti, D., Zaffaroni, N., Capelli, C., and Daidone, M. G. (2006). Breast cancer stem cells: An overview. *Eur. J. Cancer* **42,** 1219–1224.

Prockop, D. J. (1997). Marrow stromal cells as stem cells for nonhematopoietic tissues. *Science* **276,** 71–74.

Raff, A. B., Gray, A., and Kast, W. M. (2009). Prostate stem cell antigen: A prospective therapeutic and diagnostic target. *Cancer Lett.* **277,** 126–132.

Rajalingam, K., Wunder, C., Brinkmann, V., Churin, Y., Hekman, M., Sievers, C., Rapp, U. R., and Rudel, T. (2005). Prohibitin is required for Ras-induced Raf-MEK-ERK activation and epithelial cell migration. *Nat. Cell Biol.* **7,** 837–843.

Rajotte, D., Arap, W., Hagedorn, M., Koivunen, E., Pasqualini, R., and Ruoslahti, E. (1998). Molecular heterogeneity of the vascular endothelium revealed by in vivo phage display. *J. Clin. Invest.* **102,** 430–437.

Reddi, A. H., Roodman, D., Freeman, C., and Mohla, S. (2003). Mechanisms of tumor metastasis to the bone: Challenges and opportunities. *J. Bone Miner. Res.* **18,** 190–194.

Reya, T., Morrison, S. J., Clarke, M. F., and Weissman, I. L. (2001). Stem cells, cancer, and cancer stem cells. *Nature* **414,** 105–111.

Richardson, G. D., Robson, C. N., Lang, S. H., Neal, D. E., Maitland, N. J., and Collins, A. T. (2004). CD133, a novel marker for human prostatic epithelial stem cells. *J. Cell. Sci.* **117,** 3539–3545.

Rosen, J. M., and Jordan, C. T. (2009). The increasing complexity of the cancer stem cell paradigm. *Science* **324,** 1670–1673.

Rosen, E. D., and MacDougald, O. A. (2006). Adipocyte differentiation from the inside out. *Nat. Rev. Mol. Cell Biol.* **7,** 885–896.

Roth, J., Vogl, T., Sorg, C., and Sunderkotter, C. (2003). Phagocyte-specific S100 proteins: A novel group of proinflammatory molecules. *Trends Immunol.* **24,** 155–158.

Ruoslahti, E. (2002a). Drug targeting to specific vascular sites. *Drug Discov. Today* **7,** 1138–1143.

Ruoslahti, E. (2002b). Specialization of tumour vasculature. *Nat. Rev. Cancer* **2,** 83–90.

Samoylova, T. I., and Smith, B. F. (1999). Elucidation of muscle-binding peptides by phage display screening. *Muscle Nerve* **22,** 460–466.

Samoylova, T. I., Morrison, N. E., Globa, L. P., and Cox, N. R. (2006). Peptide phage display: Opportunities for development of personalized anti-cancer strategies. *Anticancer Agents Med. Chem.* **6,** 9–17.

Scanlan, M. J., Gure, A. O., Jungbluth, A. A., Old, L. J., and Chen, Y. T. (2002). Cancer/testis antigens: An expanding family of targets for cancer immunotherapy. *Immunol. Rev.* **188,** 22–32.

Schmitz, K. R., and Ferguson, K. M. (2009). Interaction of antibodies with ErbB receptor extracellular regions. *Exp. Cell Res.* **315,** 659–670.

Scott, L. J., Clarke, N. W., George, N. J., Shanks, J. H., Testa, N. G., and Lang, S. H. (2001). Interactions of human prostatic epithelial cells with bone marrow endothelium: Binding and invasion. *Br. J. Cancer* **84,** 1417–1423.

Sengle, G., Charbonneau, N. L., Ono, R. N., Sasaki, T., Alvarez, J., Keene, D. R., Bachinger, H. P., and Sakai, L. Y. (2008). Targeting of bone morphogenetic protein growth factor complexes to fibrillin. *J. Biol. Chem.* **283,** 13874–13888.

Sergeeva, A., Kolonin, M. G., Molldrem, J. J., Pasqualini, R., and Arap, W. (2006). Display technologies: Application for the discovery of drug and gene delivery agents. *Adv. Drug Deliv. Rev.* **58,** 1622–1654.

Sharma, A., and Qadri, A. (2004). Vi polysaccharide of Salmonella typhi targets the prohibitin family of molecules in intestinal epithelial cells and suppresses early inflammatory responses. *Proc. Natl. Acad. Sci. USA* **101,** 17492–17497.

Shimomura, Y., Ogawa, A., Kawada, M., Sugimoto, K., Mizoguchi, E., Shi, H. N., Pillai, S., Bhan, A. K., and Mizoguchi, A. (2008). A unique B2 B cell subset in the intestine. *J. Exp. Med.* **205,** 1343–1355.

Shmelkov, S. V., Butler, J. M., Hooper, A. T., Hormigo, A., Kushner, J., Milde, T., St Clair, R., Baljevic, M., White, I., Jin, D. K., *et al.* (2008). CD133 expression is not restricted to stem cells, and both CD133+ and CD133− metastatic colon cancer cells initiate tumors. *J. Clin. Invest.* **118,** 2111–2120.

Short, B. J., Brouard, N., and Simmons, P. J. (2009). Prospective isolation of mesenchymal stem cells from mouse compact bone. *Methods Mol. Biol.* **482,** 259–268.

Simmons, P. J., and Torok-Storb, B. (1991). Identification of stromal cell precursors in human bone marrow by a novel monoclonal antibody, STRO-1. *Blood* **78,** 55–62.

Simons, K., and Ikonen, E. (1997). Functional rafts in cell membranes. *Nature* **387,** 569–572.

Singh, S. K., Hawkins, C., Clarke, I. D., Squire, J. A., Bayani, J., Hide, T., Henkelman, R. M., Cusimano, M. D., and Dirks, P. B. (2004). Identification of human brain tumour initiating cells. *Nature* **432,** 396–401.

Sleeman, J. P., Krishnan, J., Kirkin, V., and Baumann, P. (2001). Markers for the lymphatic endothelium: In search of the holy grail? *Microsc. Res. Tech.* **55,** 61–69.

Stappenbeck, T. S., and Miyoshi, H. (2009). The role of stromal stem cells in tissue regeneration and wound repair. *Science* **324,** 1666–1669.

Steeg, P. S. (2006). Tumor metastasis: Mechanistic insights and clinical challenges. *Nat. Med.* **12,** 895–904.

Storch, J., and Thumser, A. E. (2000). The fatty acid transport function of fatty acid-binding proteins. *Biochim. Biophys. Acta* **1486,** 28–44.

Stremmel, W., Pohl, L., Ring, A., and Herrmann, T. (2001). A new concept of cellular uptake and intracellular trafficking of long-chain fatty acids. *Lipids* **36,** 981–989.

Su, X., and Abumrad, N. A. (2009). Cellular fatty acid uptake: A pathway under construction. *Trends Endocrinol. Metab.* **20,** 72–77.

Sumi, T., Ishiko, O., Yoshida, H., Hyun, Y., and Ogita, S. (2001). Involvement of angiogenesis in weight-loss in tumor-bearing and diet-restricted animals. *Int. J. Mol. Med.* **8,** 533–536.

Tang, D. G., Patrawala, L., Calhoun, T., Bhatia, B., Choy, G., Schneider-Broussard, R., and Jeter, C. (2007). Prostate cancer stem/progenitor cells: Identification, characterization, and implications. *Mol. Carcinog.* **46,** 1–14.

Tang, W., Zeve, D., Suh, J. M., Bosnakovski, D., Kyba, M., Hammer, R. E., Tallquist, M. D., and Graff, J. M. (2008). White fat progenitor cells reside in the adipose vasculature. *Science* **322,** 583–586.

Termine, J. D., Kleinman, H. K., Whitson, S. W., Conn, K. M., McGarvey, M. L., and Martin, G. R. (1981). Osteonectin, a bone-specific protein linking mineral to collagen. *Cell* **26,** 99–105.

Thedieck, C., Kalbacher, H., Kuczyk, M., Muller, G. A., Muller, C. A., and Klein, G. (2007). Cadherin-9 is a novel cell surface marker for the heterogeneous pool of renal fibroblasts. *PLoS ONE* **2**, e657.

Traktuev, D., Merfeld-Clauss, S., Li, J., Kolonin, M., Arap, W., Pasqualini, R., Johnstone, B. H., and March, K. L. (2008). A Population of multipotent CD34-positive adipose stromal cells share pericyte and mesenchymal surface markers, reside in a periendothelial location, and stabilize endothelial networks. *Circ. Res.* **102**, 77–85.

Trepel, M., Grifman, M., Weitzman, M. D., and Pasqualini, R. (2000). Molecular adaptors for vascular-targeted adenoviral gene delivery. *Hum. Gene Ther.* **11**, 1971–1981.

Valadon, P., Garnett, J. D., Testa, J. E., Bauerle, M., Oh, P., and Schnitzer, J. E. (2006). Screening phage display libraries for organ-specific vascular immunotargeting *in vivo*. *Proc. Natl. Acad. Sci. USA* **103**, 407–412.

van Beijnum, J. R., Petersen, K., and Griffioen, A. W. (2009). Tumor endothelium is characterized by a matrix remodeling signature. *Front. Biosci.* **1**, 216–225.

Voermans, C., van Hennik, P. B., and van Der Schoot, C. E. (2001). Homing of human hematopoietic stem and progenitor cells: New insights, new challenges? *J. Hematother. Stem Cell Res.* **10**, 725–738.

Vogelstein, B., and Kinzler, K. W. (2004). Cancer genes and the pathways they control. *Nat. Med.* **10**, 789–799.

Voros, G., Maquoi, E., Demeulemeester, D., Clerx, N., Collen, D., and Lijnen, H. R. (2005). Modulation of angiogenesis during adipose tissue development in murine models of obesity. *Endocrinology* **146**, 4545–4554.

Wagner, W., Wein, F., Seckinger, A., Frankhauser, M., Wirkner, U., Krause, U., Blake, J., Schwager, C., Eckstein, V., Ansorge, W., and Ho, A. D. (2005). Comparative characteristics of mesenchymal stem cells from human bone marrow, adipose tissue, and umbilical cord blood. *Exp. Hematol.* **33**, 1402–1416.

Wang, P., Mariman, E., Keijer, J., Bouwman, F., Noben, J. P., Robben, J., and Renes, J. (2004). Profiling of the secreted proteins during 3T3–L1 adipocyte differentiation leads to the identification of novel adipokines. *Cell. Mol. Life Sci.* **61**, 2405–2417.

Wang, S., Garcia, A. J., Wu, M., Lawson, D. A., Witte, O. N., and Wu, H. (2006). Pten deletion leads to the expansion of a prostatic stem/progenitor cell subpopulation and tumor initiation. *Proc. Natl. Acad. Sci. USA* **103**, 1480–1485.

Wasserman, F. (1965). The development of adipose tissue. *In* "Handbook of Physiology" (A. Renold and G. Cahill, eds.), pp. 87–100. American Physiological Society, Washington, DC.

Wels, J., Kaplan, R. N., Rafii, S., and Lyden, D. (2008). Migratory neighbors and distant invaders: Tumor-associated niche cells. *Genes Dev.* **22**, 559–574.

White, E. S., Baralle, F. E., and Muro, A. F. (2008). New insights into form and function of fibronectin splice variants. *J. Pathol.* **216**, 1–14.

Zhang, L., Hoffman, J. A., and Ruoslahti, E. (2005). Molecular profiling of heart endothelial cells. *Circ. Res.* **112**, 1601–1611.

Zhang, Y., Daquinag, A., Traktuev, D. O., Amaya, F., Simmons, P. J., March, K. L., Pasqualini, R., Arap, W., and Kolonin, M. G. (2009). White adipose tissue cells are recruited by experimental tumors and promote cancer progression in mouse models. *Cancer Res.* **69**, 5259–5266.

Zurita, A. J., Troncoso, P., Cardó-Vila, M., Logothetis, C. J., Pasqualini, R., and Arap, W. (2004). Combinatorial screenings in patients: The interleukin-11 receptor alpha as a candidate target in the progression of human prostate cancer. *Cancer Res.* **64**, 435–439.

4

Ligand-directed Cancer Gene Therapy to Angiogenic Vasculature

Wouter H. P. Driessen, Michael G. Ozawa, Wadih Arap, and Renata Pasqualini
David H. Koch Center, The University of Texas M. D. Anderson Cancer Center, Houston, Texas 77030, USA

ABSTRACT

Gene therapy strategies in cancer have remained an active area of preclinical and clinical research. One of the current limitations to successful trials is the relative transduction efficiency to produce a therapeutic effect. While intratumoral injections are the mainstay of many treatment regimens to date, this approach is hindered by hydrostatic pressures within the tumor and is not always applicable to all tumor subtypes. Vascular-targeting strategies introduce an alternative method to deliver vectors with higher local concentrations and minimization

Advances in Genetics, Vol. 67
0065-2660/09 $35.00
DOI: 10.1016/S0065-2660(09)67004-8

of systemic toxicity. Moreover, therapeutic targeting of angiogenic vasculature often leads to enhanced bystander effects, improving efficacy. While identification of functional and systemically accessible molecular targets is challenging, approaches, such as *in vivo* phage display and phage-based viral delivery vectors, provide a platform upon which vascular targeting of vectors may become a viable and translational approach. © 2009, Elsevier Inc.

I. INTRODUCTION

Cancer is a heterogeneous disease marked by aberrant cellular growth. It remains one of the leading causes of mortality in the United States and in the last several years has shown increases in incidence (National Center for Health Statistics and Centers for Disease Control and Prevention, 2006). While improvements have been made in standard treatment regimens for solid tumors, including gamma-knife surgery and radiation- and/or chemotherapy, the survival rates vary widely both between tumor types and between individual patients. For example, in the case of pancreatic tumors, the median survival is less than 6 months, despite aggressive standard therapies (Greenlee *et al.*, 2000), whereas other tumors can have overall survival rates of greater than 70% in 5 years, such as prostate cancer (National Center for Health Statistics and Centers for Disease Control and Prevention, 2006). In addition, resistance to radiation- and/or chemotherapy as well as metastatic spread for advanced tumors further complicate treatment and disease prognosis. Therefore, a need remains for newer alternative therapies that would be applicable to many, if not all, solid tumor types and that would have efficacy in a setting of advanced tumor development where genetic or epigenetic alterations in tumor cells enhance resistance.

Through early advances in molecular biology that enabled scientists to sequence and clone genes, the field of gene therapy emerged with a rationale to treat disease by replacing, manipulating, or supplementing nonfunctional genes. Numerous basic and preclinical studies lead to the first clinical trial in 1989 in which Rosenberg *et al.* (1990) used *ex vivo* gene therapy with retroviruses to treat metastatic melanoma. Enthusiasm for gene therapy strategies in cancer remains high, as nearly two-thirds of all current clinical gene therapy trials are directed against cancer (Edelstein *et al.*, 2004).

The emergent data from clinical gene therapy trials have brought to light the contribution of numerous variables for successful end results. Noted factors include gene target regulation, cell transduction efficiency, duration of gene expression, vector stability, and allowing for readministration. To optimize these factors, and thereby minimize variability, the choice of vector delivery system remains crucial. The most widely used vector remains adenovirus, with recent increased use of adeno-associated virus (AAV) and nonviral delivery systems. However, use of these approaches often necessitates intratumoral or local injection and attempts to deliver these vectors systemically have met with poor results. One of the basic tenets of systemic targeting is that the first cellular layer a circulating agent would encounter is the endothelial lining of blood vessels. The introduction of vascular targeting to gene-delivery vehicles could permit higher local concentrations for transduction, increase exposure, and minimize systemic toxicity. This review focuses on therapeutic concepts for targeted cancer gene therapy, vectors suitable for site-directed delivery, and methods to identify suitable receptors for ligand-directed delivery.

II. THERAPEUTIC CONCEPTS IN CANCER GENE THERAPY

The complexity of the tissue and tumor microenvironment permits a number of different targeting strategies toward different cell types relevant for therapy. The abundant genetic abnormalities in tumor cells present a clear target for genetic manipulation. In addition, introduction of genes into genetically stable cellular components in the tumor, such as the stroma and endothelial cells of blood vessels, provides an alternative strategy for delivery. Another approach involves stimulation of the immune system for tumor growth inhibition. The advantages and disadvantages of these methods along with current concepts for target genes are further explored in the following sections.

A. Immunomodulation

Intense study in the area of immunology over the last decade has made cancer immunobiology one of the more promising and dominant approaches in cancer gene therapy (Blankenstein et al., 1996). The goal is to stimulate a host response against the tumor by enhancing or inducing the native immune system using direct vaccination and immunization of tumor antigens.

To enhance the immunogenicity of tumors, transfection of an individual's tumor cells and autologous vaccination have emerged as successful methods for gene delivery. The tumor cells transfected with a number of candidate genes for use in this type of treatment are genes expressing costimulators of T-cell activation (e.g., CD80, CD86, and CD40) (Vesosky and Hurwitz, 2003); cytokines (e.g., interleukin-2 (IL-2), IL-3, IL-4, IL-6, IL-7, IL-10, IL-12, granulocyte-macrophage colony-stimulating factor (GM-CSF), tumor necrosis factor (TNF), and interferon-γ) to facilitate differentiation and/or activation of effector cells (Qian et al., 2006); allogeneic MHC class I proteins (e.g., transfection of HLA-B7 into the tumors of HLA-B7-negative patients) (Nabel et al., 1993); or syngeneic MHC class II proteins whose expression enables the tumor cell to present antigens to T-lymphocytes and stimulates activation of T-helper cells (Hock et al., 1995).

The introduction of these genes permits activation and targeting of the tumor cells for elimination. These strategies have led to several Phase I/II trials, including in melanoma patients, with autologous vaccination of irradiated, transduced tumor cells, and direction adenoviral delivery of antigens such as MART-1 and GP100 (J.Gene.Med, 2006).

A different approach uses immunization protocols, where gene-delivery vectors express known tumor antigens on the surface of muscle cells, dendritic cells, or T-lymphocytes. These cells in turn stimulate antigen-presenting cells or secondary stimulatory cells for activation of an immune response. An important consideration in this approach is avoidance of sensitization to nontumor cells and antigens by using selectively stimulating antigens, such as those only expressed in embryonic tissue, those protected from immune surveillance (e.g., cancer/testis antigens), or intracellular proteins (Acres et al., 2004; Gunther et al., 2005). Not surprisingly, the type of immunomodulation strategy to use is highly dependent on the goals of the treatment. Selection of the ideal treatment regimen requires consideration of a number of factors:

- *Type of immunity desired*: An antibody-mediated versus a cell-mediated responses require the stimulation of a different subset of T-lymphocytes.
- *Duration of response*: A potent short-term treatment may be desirable for the elimination of residual tumor cells, but if the goal is to prevent metastases, further growth or recurrence, a long-term response needs to be induced.
- *Condition of patients' immune response*: All strategies described above rely on functions of the patients' immune-system. If the patient is immune compromised because of his tumor or chemotherapy/radiation therapy, immunomodulation therapy may not be possible.
- *Tumor antigen*: In order for the immunization protocols with tumor antigens to be successful, the immunizing agent needs to be validated and expression might have to be verified for each individual patient.

These considerations, in addition to the type of cancer selected for treatment, are essential for the success of immunomodulating gene therapy for patients. With increases in understanding of the immune response, particularly in cancer patients, and the molecular mediators influencing activation or suppression, the future success of this gene therapy approach may improve.

B. Prodrug-converting enzymes

The concept of suicide-genes as a treatment modality was introduced nearly 20 years ago and has emerged as standalone treatment modality; gene-directed enzyme prodrug therapy (GDEPT). The concept uses inactive prodrugs that can be converted into active, cytotoxic drugs by enzymatic reactions within cells.

This converting enzyme is the delivered agent to the tumor site and subsequently expressed by cellular machinery. Site-specific delivery of the drug-converting enzyme at the tumor site results in a high local accumulation of cytotoxic drugs, mediating tumor elimination, and little-to-no accumulation of drug elsewhere. Moreover, the localized conversion of cytotoxic drugs also leads to a very potent bystander effect. As such, complete tumor eradication can be achieved with as little as 10% transduction of the tumor mass (Aghi et al., 2000; Rooseboom et al., 2004).

A widely used prototypical example is the herpes simplex virus thymidine kinase (HSV-TK) in combination with ganciclovir. Activation of HSV-TK phosphorylates ganciclovir to generate the toxic species (Eck et al., 1996; Moolten et al., 1990). This treatment strategy has been leveraged in numerous gene therapy trials including direct intratumoral injections in primary brain tumors and intraperitoneal injection for ovarian cancer patients.

Another example is the expression of bacterial cytosine deaminase as the converting enzyme in combination with systemically delivered 5-fluorocytosine (5-FC). Transfected cells convert 5-FC to 5-fluorouracil (5-FU) leading to cytotoxic effects (Crystal et al., 1997; Ohwada et al., 1996). One overriding advantage of these two prototype systems is the use of clinically ready prodrugs to generate a therapeutic effect, thereby streamlining approval for regulatory agencies and avoiding further complications and delays for clinical translation.

C. Tumor suppressor genes and antioncogenes

Although it is established that malignancy is not caused by a single protein or gene, there are dysregulations in several prominent genetic pathways that are very common in cancer. Two common dysregulations are the transcriptional activation of oncogenes or the transcriptional silencing of tumor suppressor genes. Thus, obvious strategies would be to treat tumors with these genetic alterations by replacing or overexpressing silenced suppressor genes or by silencing activated oncogenes. A clear advantage of these therapeutic strategies is the specificity for neoplastic cells, as tumor cells in principle are the cells in which the mutations would occur. Unfortunately, unlike the GDEPT strategies, there would be no bystander effect, necessitating an extremely high gene transfer efficiency within the tumor (e.g., nearly 100% for eradication).

There are a number of classes of tumor suppressor or transcriptionally silenced genes, which include proapoptotic genes (Fas-Ligand, TRAIL, and Bax; Norris et al., 2001) and genes involved in cell-cycle regulation (pRb, p16, and p21; Kang et al., 2002). However, the most widely studied tumor suppressor is p53, or the "guardian of the genome." p53 is mutated or deleted in over half of all human tumors and a single allele loss is often sufficient as a single hit in the two-hit hypothesis for tumorigenesis. Since p53 functions both in the cell cycle and apoptosis, the hypothesis that replacement or overexpression could serve as an extremely effective therapy (Levine, 1997). Indeed, an injectable recombinant human adenovirus expressing p53 (trademarked as Gendicine™) became the world's first gene therapy product approved by a governmental agency (State Food and Drug Administration of China (SFDA)) for the treatment of cancer. This was a milestone in the field of gene therapy and paves the way for further translational efforts (Peng, 2005).

Silencing activated oncogenes can be achieved using antisense-, ribozyme-, or RNAi-based therapies. Each of these silencing techniques relies on different mechanisms of action, but the net effect is blockage of mRNA translation into protein. In cancer biology, the following classes of genes have been targeted: (1) oncogenes; (2) cell-cycle regulatory genes; (3) drug-resistance genes; (4) angiogenic genes; (5) growth factor receptor genes; and (6) genes in cell signaling pathways (Lebedeva and Stein, 2001; McCaffrey *et al.*, 2002; Scanlon, 2004; Scanlon *et al.*, 1991; Singer *et al.*, 2003; Stein, 2001). These techniques are in early preclinical phases, but have progressed with great enthusiasm.

D. Antiangiogenesis

With work pioneered by the late Judah Folkman, it has become a well-known fact that tumors require a vascular supply to grow beyond a critical size. This realization introduced the field of angiogenesis in cancer biology and brought antiangiogenesis therapy as a viable new strategy to treat the disease. Antiangio-genesis treatments seek to eliminate or inhibit vascular expansion to reduce tumor burden. A number of naturally generated inhibitors of angiogenesis have been studied as a gene therapy modality, including angiostatin and endostatin (Puduvalli, 2004). Alternatively, downregulation of secreted proangiogenic fac-tors, such as VEGF or bFGF, via silencing of hypoxia inducible factor-1 alpha within the tumor have been shown to reduce tumor burden in preclinical models (Folkman, 1990; Nesbit, 2000).

Gene therapy strategies focused on endothelial cells introduce a new cellular target for exploitation and present unique advantages over therapeutic targeting of tumor cells. Despite a population density far less than tumor cells, endothelial cells, in principle, that are transduced with genes acting only within single cells would have an enhanced effect on surrounding tumor cells, akin to the bystander effect. However, this approach also intrinsically poses several unique challenges. The small cellular contribution to the tumor population creates difficulties in delivery efficiency and is likely to be very low in intratumor injections. In addition, antiangiogenic strategies also raise challenges in the selection of gene targets. Tumor cells possess genetic and epigenetic alterations that provide rational targets for intervention; however, endothelial cells lining tumor blood vessels are largely considered epigenetically stable. Therefore, selection of silencing or inhibitory products would affect normal cells throughout the body if transfected, raising the risk of unwanted side effects.

E. Combination therapies

Therapeutic gene delivery in cancer can also result in enhancement of standard treatment/regimen efficacy. A majority of modern chemotherapies do not discriminate between normal and cancer cells. Cytotoxicity to proliferating normal cells, such as hematopoietic precursor cells, becomes dose-limiting in treatment with chemotherapeutics. Thus, methods to reduce toxicity to normal cells would be a major advance in treatment regimens. For example, bone marrow depletion remains a major side effect of chemotherapy. Transfection of bone marrow cells with multidrug-resistant 1 gene enhances cellular resistance to chemotherapy and allows patients to receive higher doses of conventional agents (Culver, 1996; Huber and Margrath, 1998; Lattime and Gerson, 2001; Mickisch et al., 1992; Templeton and Lasic, 2000).

Another recently suggested approach for synergistic therapy is with the introduction of iodine transporters to the tumor cells by gene delivery. These transporters increase the uptake of radioactive iodine, and this approach demonstrated success in treatment of experimental thyroid tumors (Boelaert and Franklyn, 2003). Expansion to other tumor types has been used preclinically for therapy and imaging purposes.

F. Oncolytic viruses

This strategy makes use of replicating recombinant viruses. The underlying concept is to administer the virus intratumorally after which viral replication will take place in the transduced cells. Infected cells will ultimately be disrupted and viral progeny is released, allowing the spread of infection. It is important to achieve cancer-specific replication to limit viral replication to the site of the tumor (Vecil and Lang, 2003). This can be accomplished by (1) selective cell entry, (2) selective transcription of genes necessary for replication (tumor tissue-specific promoter), or (3) deletion of genes necessary for replication in normal cells but not in tumor cells (e.g., deletion of E1B-gene in ONYX-015).

III. VECTORS FOR LIGAND-DIRECTED GENE DELIVERY

Clinical gene therapy trials have made it clear that success is controlled by several variables, one of the more important being the gene-delivery system. This dictates the type of cell to which the therapeutic genes are transferred, the expression level of the therapeutic gene and the duration of expression. The fundamental challenge of gene delivery, originates from the fact that DNA has a charged nature, is unstable in biological environments and does not cross biological barriers such as an intact endothelium and cell or nuclear membranes. The addition of targeting ligands that bind to a unique cell-surface receptor, leads to improved and more specific gene transfer to cells expressing the targeted receptor. The challenge is to identify ligands that have a sufficiently high affinity for their targets and to identify cell-surface receptors that are either unique or display increased surface density on the targeted tissue. Another requirement is that the ligand-targeted delivery vehicle gets internalized after recognizing its target receptor. This receptor-mediated endocytosis makes sure the plasmid DNA gets delivered intracellularly.

Gene-delivery systems can be divided into two general categories: biological systems (engineered viruses) and chemical systems (lipid- and polymer-based nanoparticles) (Mah *et al.*, 2002; Thomas and Klibanov, 2003; Walther and Stein, 2000; Zhdanov *et al.*, 2002). Viral gene-delivery systems are genetically engineered nonreplicating viruses capable of infecting cells and delivering their genome containing a therapeutic gene. The viral genome can be integrated into the host genome (retrovirus, lentivirus, and the later stage of AAV), or it can exist as an episome (adenovirus and the early stage of AAV infection) (Bramson and Parks, 2003; Carter, 2003; Pages and Danos, 2003). It is generally recognized that viral vectors are the most effective gene transfer vehicles; however, chemical gene-delivery systems have provided an attractive alternative to viral vectors due to their low immunogenicity, lack of replication risk, and the relative ease to manufacture them on a large scale (Thomas and Klibanov, 2003; Zhdanov *et al.*, 2002). Moreover, the ability to incorporate targeting ligands for specific homing to target tissue with little effect on manufacturing is one of the major advantages of chemical gene-delivery systems (Anwer *et al.*, 2004; Driessen *et al.*, 2008; Wood *et al.*, 2008). Changing viral tropism has been attempted as well (Buning *et al.*, 2003; Krasnykh *et al.*, 1998; Ried *et al.*, 2002; Wickham *et al.*, 1997); however, these modifications involve alteration of viral structural proteins, and it is often problematic to inactivate the endogenous viral ligand–viral receptor interaction and replace it with a new ligand (Roelvink *et al.*, 1999).

Integration of site-specific, systemic targeting of a vector with high gene-delivery profiles would create a powerful system with wide therapeutic and diagnostic application and potentially alleviate the need for invasive procedures. We recently described the development of a new class of hybrid gene-delivery vector incorporating the genetic elements of bacterial and mammalian viruses into a single entity (Hajitou *et al.*, 2006, 2007; Soghomonyan *et al.*, 2007). We exploited the genetic elements of recombinant AAV for improved mammalian cell gene expression with elements affording site-specific targeting from bacteriophage (phage) creating a novel hybrid virus termed AAVP. In a proof-of-concept study, an AAVP targeted by an RGD-containing motif (arginine-glycine-aspartic acid) homing to alpha-v-integrins was generated carrying the HSVtk gene cassette suitable for imaging and the GDEPT treatment regimen. This vector retained target specificity for alpha-v-integrins mediated by the RGD motif while retaining high transduction efficiency *in vitro*. *In vivo*, the RGD-AAVP mediated strong accumulation within the tumor following systemic administration and strong transgene expression evident 7 days after injection. Furthermore, it was demonstrated that the clinically applicable imaging of [18]FEAU could be integrated to specifically monitor the temporal dynamics and spatial heterogeneity of transgene expression over time by positron emission tomography (PET). Lastly, we observed a robust reduction in tumor burden both in murine mammary tumors in immunocompetent mice as well as in numerous human tumor xenografts in immunocompromised animals following administration of ganciclovir (Hajitou *et al.*, 2006; Soghomonyan *et al.*, 2007). Taken together, these data introduced a novel hybrid vector that may be applied in many disease settings for targeted gene therapy.

In subsequent studies, work with targeted AAVP vectors has expanded with success employing alternative transgenes as well as multiple models of human disease. In soft tissue sarcomas, the clinical standards to determine patient response often correlates poorly with patient outcome. As a proof-of-concept, targeted AAVP vectors carrying the HSV-tk suicide gene were investigated as alternative means to assess tumor response to therapy (Hajitou *et al.*, 2008). Evaluation of transgene expression by PET imaging provided a platform to repeatedly monitor localization and magnitude of gene expression and, thereby, predict the responsiveness to therapy with ganciclovir. Similarly, targeted AAVP vectors delivering an alternative transgene, tumor necrosis factor-alpha (TNFα), have been explored in preclinical models of melanoma and, more recently, in spontaneous cancers in dogs through the Comparative Oncology Trials Consortium at the National Cancer Institute (Paoloni *et al.*, 2009; Tandle *et al.*, 2009). Finally, further study of the mechanism by which AAVP vectors targeted to the vasculature-mediated tumor therapy has implicated a heterotypic bystander killing effect. This endothelial cell–tumor cell interaction is largely mediated through intercellular gap junctions involving connexins 43 and 26 (Trepel *et al.*, 2009).

At present, much of the work involving targeted AAVP vectors has been in models of human disease. However, integration of clinically applicable PET imaging with [18]FEAU and therapy with ganciclovir suggests that rapid translation to patient populations may be imminent. Furthermore, improvements in transgene regulation through developments in tissue-specific promoters may further enhance tissue specificity and improve the therapeutic index for this vector.

IV. LIGANDS FOR TARGETING ANGIOGENIC VASCULATURE

The development of new vasculature occurs during embryonic development, normal physiological processes, and in a number of pathological diseases including most solid tumors. This coordinated, multistage process, termed angiogenesis, involves the local release of growth-promoting factors and subsequent stimulation of endothelial cells lining blood vessels. Activated endothelial cells migrate, proliferate, and invade surrounding tissues, supporting the expansion of tumor cells beyond a critical size (Folkman, 1990; Folkman *et al.*, 1989; Mustonen and Alitalo, 1995). In addition, it is well established that angiogenic endothelial cells lining tumor blood vessels are morphologically and molecularly distinct (Arap and Pasqualini, 2001; Arap *et al.*, 2002; Pasqualini and Arap, 2002; Pasqualini *et al.*, 2001, 2002). The repertoire of cell-surface molecules on angiogenic blood vessels often exist as: (i) new expression of molecules not normally present on quiescent endothelial cells, (ii) elevated levels of proteins normally found at the cell surface, or (iii) rearrangement of cell-surface molecules from luminal or abluminal surfaces. It is this differential expression pattern that suggests an opportunity for site-specific targeting of angiogenic vasculature (Ozawa *et al.*, 2008). The challenge for the field, however, is the identification and validation of systemically accessible molecules with sufficient specificity and expression to mediate targeting. A number of techniques have been applied in this effort. Genomic approaches rely on expression differences of tumor endothelium compared to normal blood vessels. St Croix *et al.* utilized microdissection combined with serial analysis of gene expression (SAGE) to identify several candidate tumor endothelial markers in human colorectal cancers (Saha *et al.*, 2001; St Croix *et al.*, 2000). This work demonstrated the feasibility of genetic analyses of cellular subpopulations, including endothelial cells, and the robust potential to identify targets. Moreover, similar studies have since ensued including generation of expressed sequence tags (ESTs) and analysis of cDNA microarrays. Once identified, the candidates must be validated not only as viable proteins but also must be localized to the cell surface and contribute to systemic targeting. Due to some of these inherent limitations to genetic screens, proteomic screenings often provide greater evidence for relevant and functionally significant targets. Beyond the derivation of protein arrays from cellular homogenates, techniques to directly profile the cell surface of endothelial cells have recently emerged, including *in vivo* screenings with systemically injected biotin derivates or two-dimensional peptide mapping (Roesli *et al.*, 2006a,b; Scheurer *et al.*, 2005). More recently, a report described proof-of-concept analyses *in silico* of bioinformatics-based identification of peptides inhibiting endothelial cell proliferation and migration (Karagiannis and Popel, 2008).

Our group has extensive experience in the identification of accessible targets on angiogenic vasculature using *in vivo* phage display (Kolonin *et al.*, 2001). Phage display is a highly versatile technology that involves genetically manipulating bacteriophage so that peptides or antibodies can be expressed on their surface (Smith and Petrenko, 1997). This strategy revealed a vascular address system that allows tissue-specific targeting of normal blood vessels and angiogenesis-related targeting of tumor blood vessels. Vascular receptors corresponding to the selected peptides have been identified in blood vessels of normal organs and in tumor blood vessels. Our strategy has shown that it is possible to shed light into selective expression of biologically relevant targets within specialized vascular beds. In the *in vivo* phage display procedure, phage capable of homing into certain organs or tumors following an intravenous injection is recovered from such phage display peptide libraries. The ability of individual peptides to target a tissue can also be analyzed by this method (Pasqualini *et al.*, 2000, 2002). In brief, phage are propagated in pilus-positive bacteria that are not lysed by the phage but rather secrete multiple of copies of phage that display a particular insert. Phage bound to a target molecule can be eluted and then amplified by growing them in host bacteria. Multiple rounds of biopanning can be performed until a population of selective binders is obtained. In addition, for a higher throughput approach and higher stringency, we have also developed an enhanced approach to phage library biopanning *in vivo* by screening a number of organs in parallel (Kolonin *et al.*, 2006b). The amino acid sequence of the recovered peptides is determined by sequencing the DNA corresponding to the insert in the phage genome. Ultimately, this approach allows circulating homing peptides to be detected in an unbiased functional assay, without any preconceived notions about the nature of their target. Aside from their carrier function for targeted gene delivery, the peptides themselves may be used as drug discovery leads for peptidomimetic drugs or for therapeutic modulation of their corresponding receptor(s), given that such receptors can be identified by biochemical or genetic approaches (Pasqualini *et al.*, 2002). Binding properties of the peptide library can also be verified for any human or mouse cell line or tissue (Kolonin *et al.*, 2006a). This biopanning strategy *in vivo* and on intact cells has several advantages. First, as opposed to purified receptors, membrane-bound proteins are more likely to preserve their functional conformation, which can be lost upon purification and immobilization outside the context of intact cells. Second, many cell-surface receptors require the cell membrane environment to function so that homo- or heterodimeric interactions may occur. Third, combinatorial approaches allow the selection of cell membrane ligands in an unbiased functional assay and without any preconceived notions about the nature of the cellular receptor repertoire; thus, unknown receptors can be targeted.

With this and related methodologies, numerous normal murine tissue-specific vascular markers and angiogenesis-related molecules in tumor blood vessels have been identified, even in human patients (Arap *et al.*, 2002). Generally, ligand–receptor pairs identified can be grouped into receptors for angiogenic proteins, adhesion molecules, metabolic receptors, extracellular matrix components, and stress-response molecules (see Table 4.1). Interestingly, some of the identified markers also serve as viral receptors, such as alpha-v-integrins (receptors for adenovirus; Wickham *et al.*, 1993), CD13/APN (a receptor for coronaviruses; Look *et al.*, 1989; Yeager *et al.*, 1992), and MMP-2 and MMP-9 (shown to be receptors for echoviruses; Pulli *et al.*, 1997). It is tempting to speculate that bacteriophage,

Table 4.1. Validated Cell-Surface Receptors and Homing Motifs Isolated by *In Vivo* Phage Display

Receptor	Localization	Homing motif	References
Receptors for angiogenic proteins			
VEGFR1; Neuropilin-1	ECs	CPQPRPLC	Giordano *et al.* (2005)
bFGFR	N.D.	MQLPLAT	Maruta *et al.* (2002)
VEGFR2	N.D.	ATWLPPR	Binetruy-Tournaire *et al.* (2000)
Adhesion molecules			
αv β3-integrin	ECs, tumor cells	CDCRGDCFC; RGD-containing moieties	Pasqualini *et al.* (1995, 1997), Temming *et al.* (2005)
αv β5-integrin	ECs, tumor cells	CMLAGWIPC CWLGEWLGC	Nie *et al.* (2008)
MCAM/MUC18	ECs, tumor cells	CLFMRLAWC	Staquicini *et al.* (2008)
VCAM-1	N.D.	VHSPNKK	Joyce *et al.* (2003), Kelly *et al.* (2005)
Extracellular matrix components			
CD13	ECs, pericytes	CNGRC	Pasqualini *et al.* (2000)
Aminopeptidase A	Pericytes, stroma	CPRECESIC	Marchio *et al.* (2004)
NG2/HMWMAA	Pericytes, tumor	GSL	Burg *et al.* (1999)
MMP-2/MMP-9	ECs, tumor cells	CTTHWGFTLC	Koivunen *et al.* (1999)
MDP	ECs (lung)	GFE	Rajotte *et al.* (1998)
Stress-response molecules			
GRP78	Tumor cells	WIFPWIQL WDLAWMFRLPVG	Arap *et al.* (2004)
HSP90	Tumor cells	CVPELGHEC	Vidal *et al.* (2004)
Miscellaneous			
IL-11R	ECs, tumor cells	CGRRAGGSC	Arap *et al.* (2002), Cardo-Vila *et al.* (2008)
CRKL	Tumor cells	YRCTLNSPFF-WEDMTHECHA	Mintz *et al.* (2009)
Prohibitin	ECs on WAT	CKGGRAKDC	Kolonin *et al.* (2004)

VEGF, vascular endothelial growth factor; bFGF, basic fibroblast growth factor; MCAM, melanoma a cell adhesion molecule; EC, endothelial cells; HMWMAA, high molecular weight melanoma-associated antigen; MMP, matrix metalloproteinase; CRKL, chicken tumor no. 10 regulator of kinase-like protein; HSP, heat shock protein; WAT, white adipose tissue; VCAM, vascular cell adhesion molecule.

which is a class of prokaryotic viruses, could use the same cellular receptors of eukaryotic viruses given a specific targeting peptide moiety. While the natural host of bacteriophage and eukaryotic virus is vastly different, the structure of the phage capsid protein provides good evidence that bacteriophage share ancestry with animal viruses. More than an evolutionary biology footnote, these findings do suggest that the receptors isolated by *in vivo* phage display will have cell internalization capability, a key feature if one wishes to utilize peptide motifs as gene therapy carriers targeted to specific cell subpopulations.

V. CONCLUSION

One of the hallmark events in cancer progression is angiogenesis. In this chapter we have described how the unique characteristics of angiogenic tumor vasculature can be exploited to deliver genes specifically and efficiently. We explored various therapeutic gene therapy strategies and methods to uncover vascular ZIP-codes of proliferating endothelium have been described. The ligand–receptor pairs discovered by such technologies can be used to target gene-delivery vehicles. We have also highlighted a new hybrid gene-delivery vector (AAVP) which has shown antitumoral efficacy in multiple animal models and tumor subtypes. In conclusion, vascular-targeting strategies for cancer gene therapy may become a new treatment paradigm to improve and enhance current therapeutic protocols.

References

Acres, B., Paul, S., Haegel-Kronenberger, H., Calmels, B., and Squiban, P. (2004). Therapeutic cancer vaccines. *Curr. Opin. Mol. Ther.* **6**(1), 40–47.

Aghi, M., Hochberg, F., and Breakefield, X. O. (2000). Prodrug activation enzymes in cancer gene therapy. *J. Gene. Med.* **2**(3), 148–164.

Anwer, K., Kao, G., Rolland, A., Driessen, W. H., and Sullivan, S. M. (2004). Peptide-mediated gene transfer of cationic lipid/plasmid DNA complexes to endothelial cells. *J. Drug Target.* **12**(4), 215–221.

Arap, W., and Pasqualini, R. (2001). The human vascular mapping project. Selection and utilization of molecules for tumor endothelial targeting. *Haemostasis* **31**(Suppl. 1), 30–31.

Arap, W., Kolonin, M. G., Trepel, M., Lahdenranta, J., Cardo-Vila, M., Giordano, R. J., Mintz, P. J., Ardelt, P. U., Yao, V. J., Vidal, C. I., Chen, L., Flamm, A., *et al.* (2002). Steps toward mapping the human vasculature by phage display. *Nat. Med.* **8**(2), 121–127.

Arap, M. A., Lahdenranta, J., Mintz, P. J., Hajitou, A., Sarkis, A. S., Arap, W., and Pasqualini, R. (2004). Cell surface expression of the stress response chaperone GRP78 enables tumor targeting by circulating ligands. *Cancer Cell* **6**(3), 275–284.

Binetruy-Tournaire, R., Demangel, C., Malavaud, B., Vassy, R., Rouyre, S., Kraemer, M., Plouet, J., Derbin, C., Perret, G., and Mazie, J. C. (2000). Identification of a peptide blocking vascular endothelial growth factor (VEGF)-mediated angiogenesis. *EMBO J.* **19**(7), 1525–1533.

Blankenstein, T., Cayeux, S., and Qin, Z. (1996). Genetic approaches to cancer immunotherapy. *Rev. Physiol. Biochem. Pharmacol.* **129**, 1–49.

Boelaert, K., and Franklyn, J. A. (2003). Sodium iodide symporter: A novel strategy to target breast, prostate, and other cancers? *Lancet* **361**(9360), 796–797.

Bramson, J. L., and Parks, R. J. (2003). Adenoviral vectors for gene delivery. *In* "Pharmaceutical Gene Delivery Systems" (A. Rolland and S. M. Sullivan, eds.), pp. 149–182. Marcel Dekker, New York.

Buning, H., Ried, M. U., Perabo, L., Gerner, F. M., Huttner, N. A., Enssle, J., and Hallek, M. (2003). Receptor targeting of adeno-associated virus vectors. *Gene Ther.* **10**(14), 1142–1151.

Burg, M. A., Pasqualini, R., Arap, W., Ruoslahti, E., and Stallcup, W. B. (1999). NG2 proteoglycan-binding peptides target tumor neovasculature. *Cancer Res.* **59**(12), 2869–2874.

Cardo-Vila, M., Zurita, A. J., Giordano, R. J., Sun, J., Rangel, R., Guzman-Rojas, L., Anobom, C. D., Valente, A. P., Almeida, F. C., Lahdenranta, J., Kolonin, M. G., Arap, W., *et al.* (2008). A ligand peptide motif selected from a cancer patient is a receptor-interacting site within human interleukin-11. *PLoS ONE* **3**(10), e3452.

Carter, B. (2003). Gene delivery technology: Adeno-associated virus. *In* "Pharmaceutical Gene Delivery Systems" (A. Rolland and S. M. Sullivan, eds.), pp. 183–214. Marcel Dekker, New York.

Crystal, R. G., Hirschowitz, E., Lieberman, M., Daly, J., Kazam, E., Henschke, C., Yankelevitz, D., Kemeny, N., Silverstein, R., Ohwada, A., Russi, T., Mastrangeli, A., *et al.* (1997). Phase I study of direct administration of a replication deficient adenovirus vector containing the E. coli cytosine deaminase gene to metastatic colon carcinoma of the liver in association with the oral administration of the pro-drug 5-fluorocytosine. *Hum. Gene Ther.* **8**(8), 985–1001.

Culver, K. W. (1996). "Gene therapy: A primer for physicians," pp. xvii, 198 p. Mary Ann Liebert, Inc, New York.

Driessen, W. H., Fujii, N., Tamamura, H., and Sullivan, S. M. (2008). Development of peptide-targeted lipoplexes to CXCR4-expressing rat glioma cells and rat proliferating endothelial cells. *Mol. Ther.* **16**(3), 516–524.

Eck, S. L., Alavi, J. B., Alavi, A., Davis, A., Hackney, D., Judy, K., Mollman, J., Phillips, P. C., Wheeldon, E. B., and Wilson, J.M (1996). Treatment of advanced CNS malignancies with the recombinant adenovirus H5.010RSVTK: A Phase I trial. *Hum. Gene Ther.* **7**(12), 1465–1482.

Edelstein, M. L., Abedi, M. R., Wixon, J., and Edelstein, R. M. (2004). Gene therapy clinical trials worldwide 1989–2004—An overview. *J. Gene Med.* **6**(6), 597–602.

Folkman, J. (1990). What is the evidence that tumors are angiogenesis dependent? *J. Natl. Cancer Inst.* **82**(1), 4–6.

Folkman, J., Watson, K., Ingber, D., and Hanahan, D. (1989). Induction of angiogenesis during the transition from hyperplasia to neoplasia. *Nature* **339**(6219), 58–61.

Giordano, R. J., Anobom, C. D., Cardo-Vila, M., Kalil, J., Valente, A. P., Pasqualini, R., Almeida, F. C., and Arap, W. (2005). Structural basis for the interaction of a vascular endothelial growth factor mimic peptide motif and its corresponding receptors. *Chem. Biol.* **12**(10), 1075–1083.

Greenlee, R. T., Murray, T., Bolden, S., and Wingo, P. A. (2000). Cancer statistics, 2000. *CA Cancer J. Clin.* **50**(1), 7–33.

Gunther, M., Wagner, E., and Ogris, M. (2005). Specific targets in tumor tissue for the delivery of therapeutic genes. *Curr. Med. Chem. Anticancer Agents* **5**(2), 157–171.

Hajitou, A., Trepel, M., Lilley, C. E., Soghomonyan, S., Alauddin, M. M., Marini, F. C., 3rd, Restel, B. H., Ozawa, M. G., Moya, C. A., Rangel, R., Sun, Y., Zaoui, K., *et al.* (2006). A hybrid vector for ligand-directed tumor targeting and molecular imaging. *Cell* **125**(2), 385–398.

Hajitou, A., Rangel, R., Trepel, M., Soghomonyan, S., Gelovani, J. G., Alauddin, M. M., Pasqualini, R., and Arap, W. (2007). Design and construction of targeted AAVP vectors for mammalian cell transduction. *Nat. Protoc.* **2**(3), 523–531.

Hajitou, A., Lev, D. C., Hannay, J. A., Korchin, B., Staquicini, F. I., Soghomonyan, S., Alauddin, M. M., Benjamin, R. S., Pollock, R. E., Gelovani, J. G., Pasqualini, R., and Arap, W. (2008). A preclinical model for predicting drug response in soft-tissue sarcoma with targeted AAVP molecular imaging. *Proc. Natl. Acad. Sci. USA* **105**(11), 4471–4476.

Hock, R. A., Reynolds, B. D., Tucker-McClung, C. L., and Kwok, W. W. (1995). Human class II major histocompatibility complex gene transfer into murine neuroblastoma leads to loss of tumorigenicity, immunity against subsequent tumor challenge, and elimination of microscopic preestablished tumors. *J. Immunother. Emphasis. Tumor. Immunol.* **17**(1), 12–18.

Huber, B. E., and Margrath, I. (1998). Gene therapy in the treatment of cancer: Progress and prospects. "Cancer: Clinical Science in Practice," pp. xi, 216 p. Cambridge University Press, Cambridge, UK; New York, NY.

J. Gene. Med. (2006). Gene therapy clinical trials worldwide. http://www.wiley.co.uk/genetherapy/clinical/.

Joyce, J. A., Laakkonen, P., Bernasconi, M., Bergers, G., Ruoslahti, E., and Hanahan, D. (2003). Stage-specific vascular markers revealed by phage display in a mouse model of pancreatic islet tumorigenesis. *Cancer Cell* **4**(5), 393–403.

Kang, Y., Ozbun, L. L., Angdisen, J., Moody, T. W., Prentice, M., Diwan, B. A., and Jakowlew, S. B. (2002). Altered expression of G1/S regulatory genes occurs early and frequently in lung carcinogenesis in transforming growth factor-beta1 heterozygous mice. *Carcinogenesis* **23**(7), 1217–1227.

Karagiannis, E. D., and Popel, A. S. (2008). A systematic methodology for proteome-wide identification of peptides inhibiting the proliferation and migration of endothelial cells. *Proc. Natl. Acad. Sci. USA* **105**(37), 13775–13780.

Kelly, K. A., Allport, J. R., Tsourkas, A., Shinde-Patil, V. R., Josephson, L., and Weissleder, R., 2005, Detection of vascular adhesion molecule-1 expression using a novel multimodal nanoparticle, *Circ Res* 96(3):327–36.

Koivunen, E., Arap, W., Valtanen, H., Rainisalo, A., Medina, O. P., Heikkila, P., Kantor, C., Gahmberg, C. G., Salo, T., Konttinen, Y. T., Sorsa, T., Ruoslahti, E., et al. (1999). Tumor targeting with a selective gelatinase inhibitor. *Nat. Biotechnol.* **17**(8), 768–774.

Kolonin, M., Pasqualini, R., and Arap, W. (2001). Molecular addresses in blood vessels as targets for therapy. *Curr. Opin. Chem. Biol.* **5**(3), 308–313.

Kolonin, M. G., Saha, P. K., Chan, L., Pasqualini, R., and Arap, W. (2004). Reversal of obesity by targeted ablation of adipose tissue. *Nat. Med.* **10**(6), 625–632.

Kolonin, M. G., Bover, L., Sun, J., Zurita, A. J., Do, K. A., Lahdenranta, J., Cardo-Vila, M., Giordano, R. J., Jaalouk, D. E., Ozawa, M. G., Moya, C. A., Souza, G. R., et al. (2006a). Ligand-directed surface profiling of human cancer cells with combinatorial peptide libraries. *Cancer Res.* **66**(1), 34–40.

Kolonin, M. G., Sun, J., Do, K. A., Vidal, C. I., Ji, Y., Baggerly, K. A., Pasqualini, R., and Arap, W. (2006b). Synchronous selection of homing peptides for multiple tissues by *in vivo* phage display. *FASEB J.* **20**(7), 979–981.

Krasnykh, V., Dmitriev, I., Mikheeva, G., Miller, C. R., Belousova, N., and Curiel, D. T. (1998). Characterization of an adenovirus vector containing a heterologous peptide epitope in the HI loop of the fiber knob. *J. Virol.* **72**(3), 1844–1852.

Lattime, E. C., and Gerson, S. L. (2001). "Gene Therapy of Cancer: Translational Approaches from Preclinical Studies to Clinical Implementation," pp. xix, 534 p, [7] p. of plates. Academic Press, San Diego; London.

Lebedeva, I., and Stein, C. A. (2001). Antisense oligonucleotides: Promise and reality. *Annu. Rev. Pharmacol. Toxicol.* **41**, 403–419.

Levine, A. J. (1997). p53, the cellular gatekeeper for growth and division. *Cell* **88**(3), 323–331.

Look, A. T., Ashmun, R. A., Shapiro, L. H., and Peiper, S. C. (1989). Human myeloid plasma membrane glycoprotein CD13 (gp150) is identical to aminopeptidase N. *J. Clin. Invest.* **83**(4), 1299–1307.

Mah, C., Byrne, B. J., and Flotte, T. R. (2002). Virus-based gene delivery systems. *Clin. Pharmacokinet.* **41**(12), 901–911.

Marchio, S., Lahdenranta, J., Schlingemann, R. O., Valdembri, D., Wesseling, P., Arap, M. A., Hajitou, A., Ozawa, M. G., Trepel, M., Giordano, R. J., Nanus, D. M., Dijkman, H. B., et al. (2004). Aminopeptidase A is a functional target in angiogenic blood vessels. *Cancer Cell* **5**(2), 151–162.

Maruta, F., Parker, A. L., Fisher, K. D., Hallissey, M. T., Ismail, T., Rowlands, D. C., Chandler, L. A., Kerr, D. J., and Seymour, L. W. (2002). Identification of FGF receptor-binding peptides for cancer gene therapy. *Cancer Gene Ther.* **9**(6), 543–552.

McCaffrey, A. P., Meuse, L., Pham, T. T., Conklin, D. S., Hannon, G. J., and Kay, M. A. (2002). RNA interference in adult mice. *Nature* **418**(6893), 38–39.

Mickisch, G. H., Aksentijevich, I., Schoenlein, P. V., Goldstein, L. J., Galski, H., Stahle, C., Sachs, D. H., Pastan, I., and Gottesman, M. M. (1992). Transplantation of bone marrow cells from transgenic mice expressing the human MDR1 gene results in long-term protection against the myelosuppressive effect of chemotherapy in mice. *Blood* **79**(4), 1087–1093.

Mintz, P. J., Cardo-Vila, M., Ozawa, M. G., Hajitou, A., Rangel, R., Guzman-Rojas, L., Christianson, D. R., Arap, M. A., Giordano, R. J., Souza, G. R., Easley, J., Salameh, A., *et al.* (2009). An unrecognized extracellular function for an intracellular adapter protein released from the cytoplasm into the tumor microenvironment. *Proc. Natl. Acad. Sci. USA* **106**(7), 2182–2187.

Moolten, F. L., Wells, J. M., Heyman, R. A., and Evans, R. M. (1990). Lymphoma regression induced by ganciclovir in mice bearing a herpes thymidine kinase transgene. *Hum. Gene Ther.* **1**(2), 125–134.

Mustonen, T., and Alitalo, K. (1995). Endothelial receptor tyrosine kinases involved in angiogenesis. *J. Cell Biol.* **129**(4), 895–898.

Nabel, G. J., Nabel, E. G., Yang, Z. Y., Fox, B. A., Plautz, G. E., Gao, X., Huang, L., Shu, S., Gordon, D., and Chang, A. E. (1993). Direct gene transfer with DNA-liposome complexes in melanoma: Expression, biologic activity, and lack of toxicity in humans. *Proc. Natl. Acad. Sci. USA* **90**(23), 11307–11311.

National Center for Health Statistics and Centers for Disease Control and Prevention (2006). US Mortality Public Use Data Tape 2003.

Nesbit, M. (2000). Abrogation of tumor vasculature using gene therapy. *Cancer Metastasis Rev.* **19**(1–2), 45–49.

Nie, J., Chang, B., Traktuev, D. O., Sun, J., March, K., Chan, L., Sage, E. H., Pasqualini, R., Arap, W., and Kolonin, M. G. (2008). IFATS collection: Combinatorial peptides identify alpha5beta1 integrin as a receptor for the matricellular protein SPARC on adipose stromal cells. *Stem Cells* **26**(10), 2735–2745.

Norris, J. S., Hyer, M. L., Voelkel-Johnson, C., Lowe, S. L., Rubinchik, S., and Dong, J. Y. (2001). The use of Fas Ligand, TRAIL and Bax in gene therapy of prostate cancer. *Curr. Gene Ther.* **1**(1), 123–136.

Ohwada, A., Hirschowitz, E. A., and Crystal, R. G. (1996). Regional delivery of an adenovirus vector containing the *Escherichia coli* cytosine deaminase gene to provide local activation of 5-fluorocytosine to suppress the growth of colon carcinoma metastatic to liver. *Hum. Gene Ther.* **7**(13), 1567–1576.

Ozawa, M. G., Zurita, A. J., Dias-Neto, E., Nunes, D. N., Sidman, R. L., Gelovani, J. G., Arap, W., and Pasqualini, R. (2008). Beyond receptor expression levels: The relevance of target accessibility in ligand-directed pharmacodelivery systems. *Trends Cardiovasc. Med.* **18**(4), 126–132.

Pages, J. C., and Danos, O. (2003). Retrovectors go forward. *In* "Pharmaceutical Gene Delivery Systerms" (A. Rolland and S. M. Sullivan, eds.), pp. 215–244. Marcel Dekker, New York.

Paoloni, M. C., Tandle, A., Mazcko, C., Hanna, E., Kachala, S., Leblanc, A., Newman, S., Vail, D., Henry, C., Thamm, D., Sorenmo, K., Hajitou, A., *et al.* (2009). Launching a novel preclinical infrastructure: Comparative oncology trials consortium directed therapeutic targeting of TNFalpha to cancer vasculature. *PLoS ONE* **4**(3), e4972.

Pasqualini, R., and Arap, W. (2002). Profiling the molecular diversity of blood vessels. *Cold Spring Harb. Symp. Quant. Biol.* **67,** 223–225.

Pasqualini, R., Koivunen, E., and Ruoslahti, E. (1995). A peptide isolated from phage display libraries is a structural and functional mimic of an RGD-binding site on integrins. *J. Cell Biol.* **130**(5), 1189–1196.

Pasqualini, R., Koivunen, E., and Ruoslahti, E. (1997). Alpha v integrins as receptors for tumor targeting by circulating ligands. *Nat. Biotechnol.* **15**(6), 542–546.

Pasqualini, R., Koivunen, E., Kain, R., Lahdenranta, J., Sakamoto, M., Stryhn, A., Ashmun, R. A., Shapiro, L. H., Arap, W., and Ruoslahti, E. (2000). Aminopeptidase N is a receptor for tumor-homing peptides and a target for inhibiting angiogenesis. *Cancer Res.* **60**(3), 722–727.

Pasqualini, R., McDonald, D. M., and Arap, W. (2001). Vascular targeting and antigen presentation. *Nat. Immunol.* **2**(7), 567–568.

Pasqualini, R., Arap, W., and McDonald, D. M. (2002). Probing the structural and molecular diversity of tumor vasculature. *Trends Mol. Med.* **8**(12), 563–571.

Peng, Z. (2005). Current status of gendicine in China: Recombinant human Ad-p53 agent for treatment of cancers. *Hum. Gene Ther.* **16**(9), 1016–1027.

Puduvalli, V. K. (2004). Inhibition of angiogenesis as a therapeutic strategy against brain tumors. *Cancer Treat. Res.* **117**, 307–336.

Pulli, T., Koivunen, E., and Hyypia, T. (1997). Cell-surface interactions of echovirus 22. *J. Biol. Chem.* **272**(34), 21176–21180.

Qian, C., Liu, X. Y., and Prieto, J. (2006). Therapy of cancer by cytokines mediated by gene therapy approach. *Cell Res.* **16**(2), 182–188.

Rajotte, D., Arap, W., Hagedorn, M., Koivunen, E., Pasqualini, R., and Ruoslahti, E. (1998). Molecular heterogeneity of the vascular endothelium revealed by *in vivo* phage display. *J. Clin. Invest.* **102**(2), 430–437.

Ried, M. U., Girod, A., Leike, K., Buning, H., and Hallek, M. (2002). Adeno-associated virus capsids displaying immunoglobulin-binding domains permit antibody-mediated vector retargeting to specific cell surface receptors. *J. Virol.* **76**(9), 4559–4566.

Roelvink, P. W., Mi Lee, G., Einfeld, D. A., Kovesdi, I., and Wickham, T. J. (1999). Identification of a conserved receptor-binding site on the fiber proteins of CAR-recognizing adenoviridae. *Science* **286**(5444), 1568–1571.

Roesli, C., Elia, G., and Neri, D. (2006a). Two-dimensional mass spectrometric mapping. *Curr. Opin. Chem. Biol.* **10**(1), 35–41.

Roesli, C., Neri, D., and Rybak, J. N. (2006b). *In vivo* protein biotinylation and sample preparation for the proteomic identification of organ- and disease-specific antigens accessible from the vasculature. *Nat. Protoc.* **1**(1), 192–199.

Rooseboom, M., Commandeur, J. N., and Vermeulen, N. P. (2004). Enzyme-catalyzed activation of anticancer prodrugs. *Pharmacol. Rev.* **56**(1), 53–102.

Rosenberg, S. A., Aebersold, P., Cornetta, K., Kasid, A., Morgan, R. A., Moen, R., Karson, E. M., Lotze, M. T., Yang, J. C., Topalian, S. L., Merino, M. J., and Kulver, K. (1990). Gene transfer into humans—Immunotherapy of patients with advanced melanoma, using tumor-infiltrating lymphocytes modified by retroviral gene transduction. *N. Engl. J. Med.* **323**(9), 570–578.

Saha, S., Bardelli, A., Buckhaults, P., Velculescu, V. E., Rago, C., St Croix, B., Romans, K. E., Choti, M. A., Lengauer, C., Kinzler, K. W., and Vogelstein, B. (2001). A phosphatase associated with metastasis of colorectal cancer. *Science* **294**(5545), 1343–1346.

Scanlon, K. J. (2004). Anti-genes: siRNA, ribozymes and antisense. *Curr. Pharm. Biotechnol.* **5**(5), 415–420.

Scanlon, K. J., Jiao, L., Funato, T., Wang, W., Tone, T., Rossi, J. J., and Kashani-Sabet, M. (1991). Ribozyme-mediated cleavage of c-fos mRNA reduces gene expression of DNA synthesis enzymes and metallothionein. *Proc. Natl. Acad. Sci. USA* **88**(23), 10591–10595.

Scheurer, S. B., Rybak, J. N., Roesli, C., Brunisholz, R. A., Potthast, F., Schlapbach, R., Neri, D., and Elia, G. (2005). Identification and relative quantification of membrane proteins by surface biotinylation and two-dimensional peptide mapping. *Proteomics* **5**(11), 2718–2728.

Singer, C. A., Baker, K. J., McCaffrey, A., AuCoin, D. P., Dechert, M. A., and Gerthoffer, W. T. (2003). p38 MAPK and NF-kappaB mediate COX-2 expression in human airway myocytes. *Am. J. Physiol. Lung Cell Mol. Physiol.* **285**(5), L1087–L1098.

Smith, G. P., and Petrenko, V. A. (1997). Phage display. *Chem. Rev.* **97**(2), 391–410.

Soghomonyan, S., Hajitou, A., Rangel, R., Trepel, M., Pasqualini, R., Arap, W., Gelovani, J. G., and Alauddin, M. M. (2007). Molecular PET imaging of HSV1-tk reporter gene expression using [18F] FEAU. *Nat. Protoc.* **2**(2), 416–423.

Staquicini, F. I., Tandle, A., Libutti, S. K., Sun, J., Zigler, M., Bar-Eli, M., Aliperti, F., Perez, E. C., Gershenwald, J. E., Mariano, M., Pasqualini, R., Arap, W., *et al.* (2008). A subset of host B lymphocytes controls melanoma metastasis through a melanoma cell adhesion molecule/MUC18-dependent interaction: Evidence from mice and humans. *Cancer Res.* **68**(20), 8419–8428.

St Croix, B., Rago, C., Velculescu, V., Traverso, G., Romans, K. E., Montgomery, E., Lal, A., Riggins, G. J., Lengauer, C., Vogelstein, B., and Kinzler, K. W. (2000). Genes expressed in human tumor endothelium. *Science* **289**(5482), 1197–1202.

Stein, C. A. (2001). The experimental use of antisense oligonucleotides: A guide for the perplexed. *J. Clin. Invest.* **108**(5), 641–644.

Tandle, A., Hanna, E., Lorang, D., Hajitou, A., Moya, C. A., Pasqualini, R., Arap, W., Adem, A., Starker, E., Hewitt, S., and Libutti, S. K. (2009). Tumor vasculature-targeted delivery of tumor necrosis factor-alpha. *Cancer* **115**(1), 128–139.

Temming, K., Schiffelers, R. M., Molema, G., and Kok, R. J. (2005). RGD-based strategies for selective delivery of therapeutics and imaging agents to the tumour vasculature. *Drug Resist. Updat.* **8**(6), 381–402.

Templeton, N. S., and Lasic, D. D. (2000). "Gene Therapy: Therapeutic Mechanisms and Strategies," pp. xv, 584 p. Marcel Dekker, New York.

Thomas, M., and Klibanov, A. M. (2003). Non-viral gene therapy: Polycation-mediated DNA delivery. *Appl. Microbiol. Biotechnol.* **62**(1), 27–34.

Trepel, M., Stoneham, C. A., Eleftherohorinou, H., Mazarakis, N. D., Pasqualini, R., Arap, W., and Hajitou, A. (2009). A heterotypic bystander effect for tumor cell killing after adeno-associated virus/phage-mediated, vascular-targeted suicide gene transfer. *Mol. Cancer Ther.* **8**(8), 2383–2391.

Vecil, G. G., and Lang, F. F. (2003). Clinical trials of adenoviruses in brain tumors: A review of Ad-p53 and oncolytic adenoviruses. *J. Neurooncol.* **65**(3), 237–246.

Vesosky, B., and Hurwitz, A. A. (2003). Modulation of costimulation to enhance tumor immunity. *Cancer Immunol. Immunother.* **52**(11), 663–669.

Vidal, C. I., Mintz, P. J., Lu, K., Ellis, L. M., Manenti, L., Giavazzi, R., Gershenson, D. M., Broaddus, R., Liu, J., Arap, W., and Pasqualini, R. (2004). An HSP90-mimic peptide revealed by fingerprinting the pool of antibodies from ovarian cancer patients. *Oncogene* **23**(55), 8859–8867.

Walther, W., and Stein, U. (2000). Viral vectors for gene transfer: A review of their use in the treatment of human diseases. *Drugs* **60**(2), 249–271.

Wickham, T. J., Mathias, P., Cheresh, D. A., and Nemerow, G. R. (1993). Integrins alpha v beta 3 and alpha v beta 5 promote adenovirus internalization but not virus attachment. *Cell* **73**(2), 309–319.

Wickham, T. J., Tzeng, E., Shears, L. L., 2nd, Roelvink, P. W., Li, Y., Lee, G. M., Brough, D. E., Lizonova, A., and Kovesdi, I. (1997). Increased *in vitro* and *in vivo* gene transfer by adenovirus vectors containing chimeric fiber proteins. *J. Virol.* **71**(11), 8221–8229.

Wood, K. C., Azarin, S. M., Arap, W., Pasqualini, R., Langer, R., and Hammond, P. T. (2008). Tumor-targeted gene delivery using molecularly engineered hybrid polymers functionalized with a tumor-homing peptide. *Bioconjug. Chem.* **19**(2), 403–405.

Yeager, C. L., Ashmun, R. A., Williams, R. K., Cardellichio, C. B., Shapiro, L. H., Look, A. T., and Holmes, K. V. (1992). Human aminopeptidase N is a receptor for human coronavirus 229E. *Nature* **357**(6377), 420–422.

Zhdanov, R. I., Podobed, O. V., and Vlassov, V. V. (2002). Cationic lipid-DNA complexes-lipoplexes-for gene transfer and therapy. *Bioelectrochemistry* **58**(1), 53–64.

Index

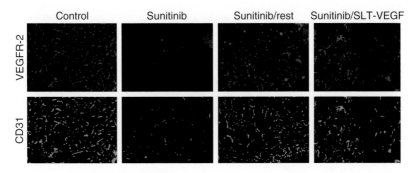

Chapter 1, Figure 1.6 (See Page 21 of this volume).

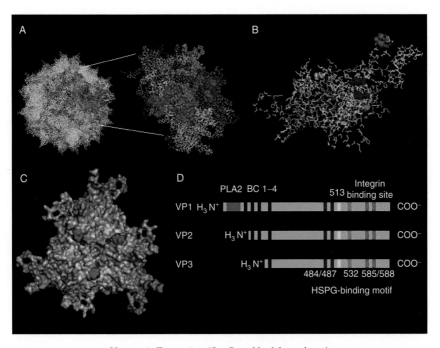

Chapter 2, Figure 2.1 (See Page 32 of this volume).

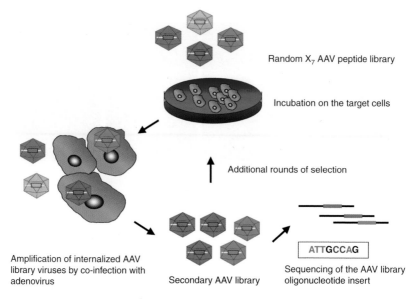

Random X₇ AAV peptide library

Incubation on the target cells

Additional rounds of selection

Amplification of internalized AAV
library viruses by co-infection with
adenovirus

Secondary AAV library

Sequencing of the AAV library
oligonucleotide insert

ATTGCCAG

Chapter 2, Figure 2.2 (See Page 44 of this volume).

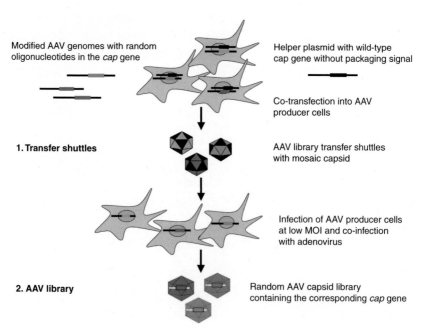

Modified AAV genomes with random
oligonucleotides in the *cap* gene

Helper plasmid with wild-type
cap gene without packaging signal

Co-transfection into AAV
producer cells

1. Transfer shuttles

AAV library transfer shuttles
with mosaic capsid

Infection of AAV producer cells
at low MOI and co-infection
with adenovirus

2. AAV library

Random AAV capsid library
containing the corresponding *cap* gene

Chapter 2, Figure 2.3 (See Page 45 of this volume).

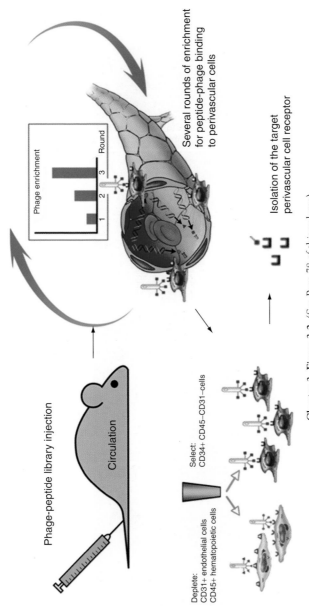

Chapter 3, Figure 3.2 (See Page 78 of this volume).

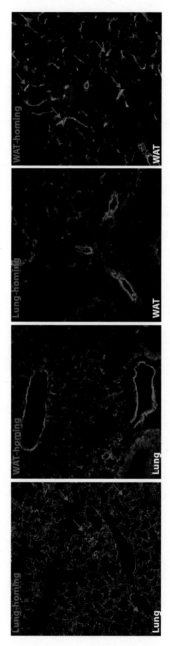

Chapter 3, Figure 3.3 (See Page 79 of this volume).

DATE DUE